3rd International Biannual Conference "Electron Beam Welding and Related Technologies" (EBW2019)

IOP Conference Series: Materials Science and Engineering
Volume 759

Moscow, Russia
12 - 15 November 2019

Editor:

Viktor Dragunov

ISBN: 978-1-7138-2661-3
ISSN: 1757-8981

Printed from e-media with permission by:

Curran Associates, Inc.
57 Morehouse Lane
Red Hook, NY 12571

Some format issues inherent in the e-media version may also appear in this print version.

Institute of Physics
Dirac House, Temple Back
Bristol BS1 6BE UK

Phone: 44 1 17 929 7481
Fax: 44 1 17 920 0979

techtracking@iop.org

Additional copies of this publication are available from:

Curran Associates, Inc.
57 Morehouse Lane
Red Hook, NY 12571 USA
Phone: 845-758-0400
Fax: 845-758-2633
Email: curran@proceedings.com
Web: www.proceedings.com

TABLE OF CONTENTS

EXPERIENCE OF FORMING COMBINED LOW-ENERGY ELECTRON-ION BEAMS IN PLASMA SOURCES OF CHARGED PARTICLES 1
D A Antonovich, V A Gruzdev

EVALUATION OF INFLUENCE OF TECHNOLOGICAL PARAMETERS ON WIDTH OF STRENGTHENED LAYER IN PLASMA SURFACE HARDENING OF STRUCTURAL STEELS 8
A E Balanovsky, V E Gozbenko, S K Kargapol'Tsev, A I Karlina, Yu I Karlina

NON-VACUUM ELECTRON BEAM WELDING AND CUTTING OF CUPPER 13
T Hassel, A Beniyash, G Klimov

EFFECTS OF LASER TREATMENT MODES ON THE PARAMETERS OF THE HARDENING ZONES AND THE WEAR RESISTANCE OF STEEL 40CR 23
V P Biryukov, V N Petrovsky, D V Panov, D V Ushakov

BEAM DIAGNOSTICS ON THE ELECTRON BEAM WELDING INSTALLATION AT BUDKER INSTITUTE OF NUCLEAR PHYSICS 29
K A Blokhina, T A Devyataykina, A M Medvedev, Yu I Semenov, M M Sizov, A A Starostenko, A S Tsyganov

ELECTRON-BEAM WELDING OF LARGE THICKNESS PARTS IN A NARROW GAP 37
I S Chulkov, A L Goncharov, A V Chulkova

CREATION OF CORROSION RESISTANT COATINGS ON DUCTILE IRON BY RE-MELTING FLAME SPRAYED LAYERS USING LASER AND ELECTRON BEAM 43
D Drimal, F Kolenic, M Simek

RESEARCH OF THERMOELECTRIC EFFECTS AND THEIR INFLUENCE ON ELECTRON BEAM IN THE PROCESS OF WELDING OF DISSIMILAR STEELS 53
A L Goncharov, A P Sliva, I A Kharitonov, A V Chulkova, E V Terentyev

RESEARCHES OF METAL TEXTURE AFTER FRICTION STIR WELDING 60
V E Gozbenko, A E Balanovsky, S K Kargapol'Tsev, Yu I Karlina, A I Karlina

RESEARCH OF MECHANICAL PROPERTIES AND CHEMICAL COMPOSITION OF SAMPLES MADE OF ALUMINIUM ALLOY AMG6 AND OBTAINED BY ELECTRON BEAM ADDITIVE MANUFACTURING WITH WIRE FEEDING 66
A V Gudenko, G V Grigoriev, A P Sliva, R Yu Agafonov, A N Kasitsyn

POSSIBILITY OF OBTAINING COMPLEX FORM DETAILS USING ADDITIVE SURFACE TECHNOLOGY 72
S K Kargapol'Tsev, A E Balanovsky, V E Gozbenko, Yu I Karlina, A I Karlina

CHALLENGES IN PATH PLANNING OF HIGH ENERGY DENSITY BEAMS FOR ADDITIVE MANUFACTURING 77
K P Karunakaran, Seema Negi, Arun Nambolan, Ashik Patel, Yogesh Patil, Milan Pandya, Sajan Kapil, Dmitriy Trushnikov, Michael Sealy, Alain Bernard

CONTROL OF ELECTRON BEAM WIELDING PARAMETERS BASED ON THE GAP SCANNING SYSTEM DATA DURING THE WELDING PROCESS 86
A N Kasitsyn, R Yu Agafonov, A V Leonov, G V Grigoriev

STUDY ON FORMATION OF ALUMINUM ALLOY THIN WALL PRODUCED WITH WAAM METHOD UNDER VARIOUS THERMAL CONDITIONS 92
A V Kisarev, N V Kobernik

ANALYSIS OF THE AMPLITUDE-TIME PARAMETERS OF CURRENT PULSES IN A PLASMA DURING LASER BEAM WELDING 99
I Yu Letyagin, V Ya Belenkiy, D N Trushnikov, S Pang

EXPRESS EVALUATION OF WELDED JOINTS COOL RESISTANCE 106
V M Matyunin, A Yu Marchenkov, M A Karimbekov, A N Demidov, P V Volkov, N Abusaif, E I Geraskin, D A Zhgut

MODIFICATION OF THE SURFACE OF STEEL 3KH2V8F BY APPLICATION B-AL-COATINGS BY METHODS OF THERMAL-CHEMICAL TREATMENT AND ELECTRON-BEAM PROCESSING 112
U L Mishigdorzhiyn, N S Ulakhanov, A S Milonov, A P Semenov

INFLUENCE OF TECHNOLOGICAL PARAMETERS OF SPRAYING ON MECHANICAL PROPERTIES OF GAS-DYNAMIC COATINGS 119
G V Moskvitin, M S Pugachev

EFFECT OF MOLYBDENUM ON THE HOT-TEARING SUSCEPTIBILITY OF THE NI-CR-MO-B SYSTEM 126
A S Nazarko, R L Plomoyalo, D V Dmitrenko

DEVELOPMENT OF METHODS FOR FORECASTING THE MICROSTRUCTURE OF THE WELDED JOINT WITH REFERENCE TO THE TEMPERATURE-TIME MODES OF ELECTRON-BEAM WELDING 132
T V Olshanskaya, E M Fedoseeva

NUMERICAL MODEL OF A HOLLOW CATHODE ARC DISCHARGE FORMATION IN VACUUM 139
G L Permyakov, D N Trushnikov, M K Mindibaev, E G Koleva

COMPUTER SIMULATION OF METALS IONIZATION PROCESS DURING ELECTRON-BEAM WELDING 144
R V Rodyakina, A V Shcherbakov, D A Gaponova

THRESHOLD CONDITIONS FOR THERMOCAPILLARY TRANSITION TO DEEP PENETRATION MODE IN SELECTIVE LASER MELTING OF METAL POWDER BED 150
R D Seidgazov, F Kh Mirzade

HEAT TREATMENT MODE OPTIMIZATION OF DISSIMILAR WELD JOINTS OF EP517 STEEL TO 36NKHTYU ALLOY 160
E V Terentyev, A Yu Marchenkov, A L Goncharov, A P Sliva, K T Borodavkina

Author Index

PREFACE

The 3rd International biannual conference "Electron Beam Welding and Related Technologies (EBW2019)" was held at National Research University "Moscow Power Engineering Institute (MPEI)" on November 12-15 2019 in Moscow, Russia. The host university MPEI is one of the leading academic institutions in Europe that conducts research in heat and electricity generation, electrotechnical instrumentation design and power engineering machinery building.

The conference is dedicated to research, innovations and novel engineering solutions in electron beam welding (EBW), metals heat treatment, and electrotechnical engineering. More than 50 reports were presented in various areas of EBW technologies, welding materials science, as well as quality control of welded joints. Reports subjects included the following areas: modeling of EBW processes (thermal fields, heat and mass transfer, electron beam energy, metal ionization, etc.), control of EBW parameters and welding equipment, additive manufacturing, diagnostics of mechanical and service properties of welded joints, electron beam welding equipment.

The target audience of the event was not only academics and students but also industry experts and state officers. The conference was attended by 150 participants from technical universities, research and production companies, factories and power plants. The showcases of the event are highly recognizable among the industrial community not only in Russia but all over the continent. This year the events were attended by delegates from such countries as Belarus, Bulgaria, Germany, China, India, Syria, Mongolia, Pakistan, Slovakia, France, and Israel.

This distinguished volume contains the most significant and novel academic research papers presented at the conference and recommended by the organizing committee. All the manuscripts went through a thorough editorial process and were peer-reviewed by two invited referees.

On behalf of the Organizing Committee, we express our gratitude to the sponsors of the conference for financial and informational support. It was a great honor for the editors to work with all the authors, reviewers and organizers.

With the best regards,
Viktor Dragunov,
The Organizing committee Chairman EBW2019 conference,
Science editor, professor of MPEI

IOP Conf. Series: Materials Science and Engineering **759** (2020) 011002 doi:10.1088/1757-899X/759/1/011002

Peer review statement

All papers published in this volume of *IOP Conference Series: Materials Science and Engineering* have been peer reviewed through processes administered by the proceedings Editors. Reviews were conducted by expert referees to the professional and scientific standards expected of a proceedings journal published by IOP Publishing.

Experience of forming combined low-energy electron-ion beams in plasma sources of charged particles

D A Antonovich, V A Gruzdev

Polotsk State University, Blokhin Street 29, Novopolotsk, 211440, Belarus,
E-mail: d.antonovich@psu.by

Abstract. The paper shows the possibility of forming combined electron-ion beams in a single multi-discharge structure that does not contain incandescent elements. A design of the electrode structure of an electron-ion source is proposed, which consists of two «Penning type» discharge cells connected in series (along the axis). It is shown that the interconnection of separately controlled discharges in the structure increases the degree of gas ionization at reduced pressure, as well as the formation of double electric layers in the plasma, which ensure the formation of combined ion-electron flows in a single structure. This is ensured by creating conditions for the drift of the electron beam through the entire part of the electrode structure, which ensures the formation of the ion current of the source, and helps to increase the degree of ionization of the gas in this region. In addition, the deceleration of the electron beam in the ion acceleration gap ensures the return of electrons, which have lost part of their energy to gas ionization, into the region of the formation of the plasma emitting ions. This contributes to an increase in the density of the ion emission current. Some experimental parameters of the beams formed by the developed structure are presented.

1. Introduction

One of the mechanisms for increasing the efficiency of ion-plasma technologies is partial or full compensation of the positive ion charge in the stream or on the treated surface. For this purpose, as a rule, additional emitting systems are used that create compensating electron flows into the discharge space, the accelerating gap, or onto the surface to be treated. The most widespread in high-power technological magnetron systems are high-current solid-state incandescent emitters of electrons, which provide stable burning of the discharge or compensation of the space charge in the beam. Under conditions of intense ion fluxes, the resource of such emitters is limited due to intensive ion bombardment, and the glow emitters themselves are characterized by high energy consumption and low efficiency. Therefore, the search non-incandescent compensatory electron emitters remain relevant.

It is known to use hollow cathode discharge systems as compensating sources [1]. However, in these systems it is necessary to coordinate control systems for the parameters formed in the independent sources of ion and electron fluxes; it is necessary to take into account the mutual influence of the magnetic fields of these sources to ensure high efficiency of the process.

At the same time, low-energy electron beams can be used for thermal exposure of the substrate to intensify plasma-chemical processes and increase the speed of diffusion processes in the substrate during coating formation. Therefore, for a number of technologies, it may be of interest not only the effect of partially or fully compensated ion beams, but also the alternating (thermophysical electronic and modifying ion) exposure to beams of charged particles of both charge signs.

To realize this effect, plasma sources of charged particles [1–4] that do not have incandescent elements are a promising tool. However, despite the fact that plasma is the emitter of charged particles in such systems, and when the polarity of the accelerating voltage is changed, beams of both signs can be formed, it is difficult to simultaneously achieve high emission efficiency in such systems, since the conditions for the emission of ions and electrons from plasma differ significantly.

It was shown [5–8] that since the conditions for the formation of ion beams in known magnetron-type systems are close, primarily in pressure, to the conditions for the formation of electron beams in systems with a plasma emitter, it is possible to create combined sources that ensure the formation of both electronic and ion beams [9].

In this paper, the possibility of the formation of combined electron-ion beams in a single multi-discharge structure that does not contain incandescent elements is shown. A design of the electrode structure of an electron-ion source is proposed, consisting of two «Penning-type» [10] gas-discharge cells connected in series (along the axis). It is shown that the interconnection of separately controlled discharges in the structure increases the degree of gas ionization at reduced pressure, as well as the formation of double electric layers in the plasma, which ensure the formation of combined ion-electron flows in a single structure. This is ensured by creating conditions for the drift of the electron beam through the entire part of the electrode structure, which ensures the formation of the ion current of the source, and helps to increase the degree of gas ionization in this region. In addition, the deceleration of the electron beam in the ion acceleration gap ensures the return of electrons, which have lost part of their energy to gas ionization, into the region of the formation of the plasma emitting ions. This contributes to an increase in the density of the ion emission current. The possibility of separate regulation of the accelerating voltages of electrons and ions in the developed structure provides regulation in a wide range of ratios of the energies of electrons and ions, as well as their current densities in the electron-ion beam. This broadens the range of possible technological applications of the electron-ion beam source. The results of the studies indicate the promise of using the developed structure for the design of technological sources of combined electron-ion beams.

2. Experimental models of sources of charged particles

Designed layout of electron-ion source, the appearance and outline of the electrode structure of which is shown in Figure 1, is a generator emitting plasma formed in the volume limited by the inner surfaces of the cathodes 2 and 4, the anode 3 and the emitter electrode 5 (discharge chamber I). Electrodes 6 and 7 form an electron acceleration gap where a plasma surface is formed that emits electrons. Electrodes 8-12 form a gas-discharge structure forming plasma, which is a source of atomizing ions. This structure consists of two «Penning type» discharge cells (II and III) connected in series (along the axis). Elements of this structure 9 and 11 are the anodes of the discharge cells; elements 8, 10 and 12 - cathodes, which are simultaneously pole tips of permanent magnets, providing electron oscillation between the cathodes of this (second) gas-discharge structure.

a)

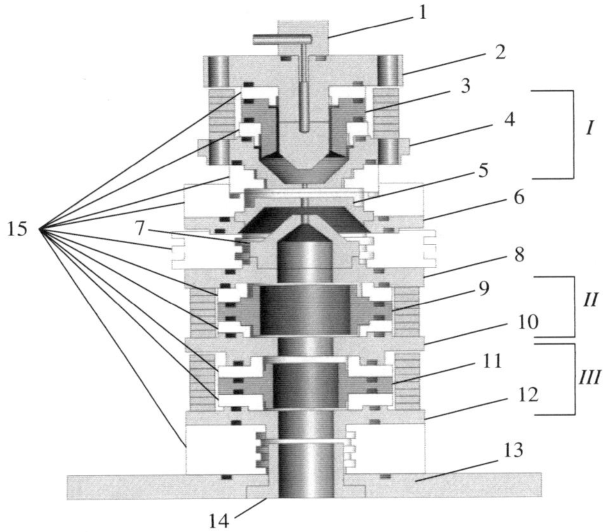

b)

Fig. 1 - Appearance and internal structure of the developed layout electron-ion source with crossed E × H fields: (1) fitting for plasma gas inlet, (2) internal cathode, (3) the main anode, (4) external cathode, (5) emitter electrode, (6) auxiliary anode, (7) accelerating electrode, (8, 10, 12) cathodes, (9, 11) anodes, (13) flange for mounting the structure on the working chamber, (14) matching electrode, and (15) insulators; I, II, III – Areas of discharge chambers

At the same time, the magnetic field generated by the cathodes 8, 10 and 12 forms a magnetic focusing system for an accelerated electron beam propagating along the axis of this (second) gas-discharge structure until the ion-electron beam exits the source into the process chamber. A voltage is applied between the electrodes 12 and 14, accelerating the ions to the energy of the atomizing ions required by the technology. At the same time, in this gap (between the electrodes 12 and 14), the beam of electrons accelerated in the gap between the electrodes 6 and 7 is decelerated. The plasma surface emitting ions between the electrodes 12 and 14 determines the trajectories of both ions and electrons in the drift space of the electron-ion beam to the sputtering target, and hence determines the distribution of the ion and electron current densities over the target surface. Figure 2 shows a diagram of the connection of power supplies to the layout electrodes.

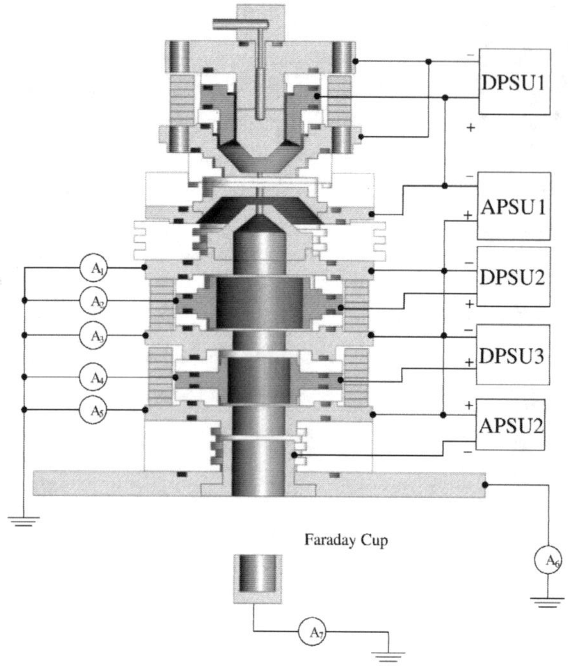

Fig. 2 - Diagram of the connection of the electrodes of the discharge chambers of the electron-ion source based on discharge in crossed E × H fields: DPSU1, DPSU2, DPSU3 – power supply unit of the corresponding discharge (I, II, or III camera) with discharge voltage U_d up to 1000 V and current I_d up to 1.2 A; APSU - accelerating power supply unit with voltage U_{ac} up to 5 kV and current up to 0.5 A; APSU 1, electron acceleration unit; APSU 2, ion acceleration unit

3. Results and Discussion

The initiation of a discharge in two discharge chambers (I and II - Fig. 1) in the electron extraction mode indicates the prospect of using similar structures as prototypes of high-plasma electron sources. The following mechanism of operation in this mode is assumed: the electron beam formed in the I chamber (Fig. 1), after acceleration, gets into the structure formed by the electrodes of the chamber II (Fig. 1), propagates along the axis in this structure and initiates a low-pressure discharge in which a plasma that emits ions is formed. Ions propagate into the upper structure, increasing the ionization of the gas in the region of electron extraction, increasing the density of the emission current and partially compensate for the space charge of the accelerating gap, which generally leads to an increase in the source perveance. This is evidenced by the current-voltage characteristics of extraction in the presence and absence of discharge initiation in the discharge chamber II (Fig. 1) of the source layout shown in Figure 3 and Figure 4. The presented characteristics indicate a weak influence on the perveance of the source of variation in the working range of the gas inlet (pressure in the discharge chamber) and the discharge current. From the presented characteristics it is seen that the determining effect on the source perveance is exerted by the presence of an additional discharge in chamber II (Fig. 1).

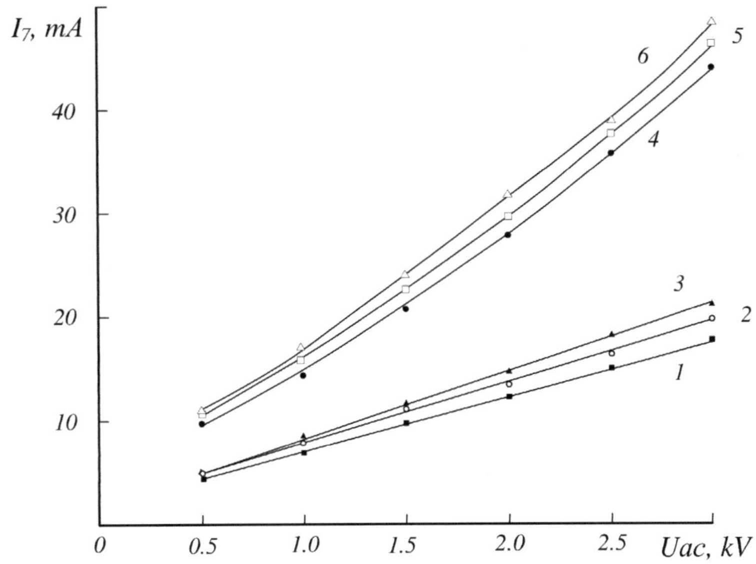

Fig. 3 - Current I_7 (into the Faraday cup) in the absence (1-3) and the presence of (2-4) initiation discharge in discharge chamber II: I_d in chamber I (Fig. 1) is 200 mA, the discharge voltage of the discharge is 420 V, I_d in chamber II (Fig. 1) is 180 mA, and the discharge voltage of the discharge is 410 V

Gas inlet Q: 1, 4 – 1,1 mPa·m^3/s; 2, 5 – 2,2 mPa·m^3/s; 3, 6 – 3,5 mPa·m^3/s

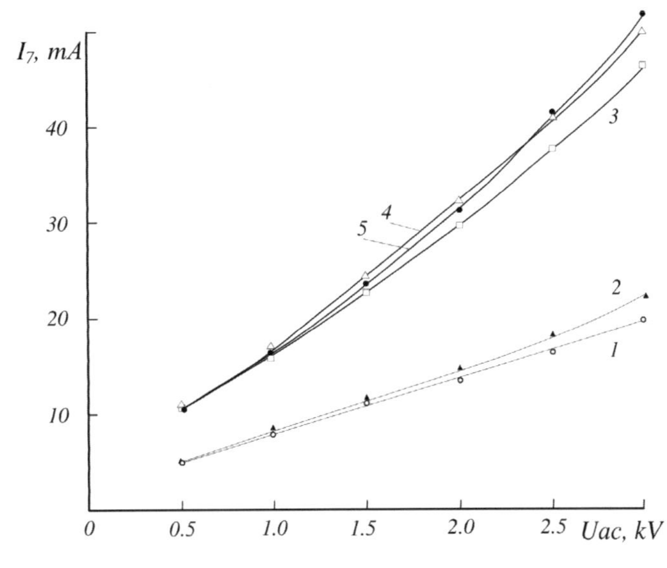

Fig. 4 - Current I_7 (into the Faraday cylinder) in the absence (1-2) and the presence of (3-5) initiation discharge in discharge chamber II for different discharge current into the chambers I and II: 1-5 – gas inlet Q – 2,2 mPa·m^3/s;

3, 4 – I_d in chamber II (fig. 1) 180 mA, the discharge voltage 410 V;

5 – I_d in chamber II (fig. 1) 210 mA, the discharge voltage 430 V;

I_d in chamber II (fig. 1): 1, 3, 5 – 200 mA, the discharge voltage 420 V;

2, 4 – I_d in chamber II (fig. 1) 240 mA, the discharge voltage 450 V

Figure 5 shows the current-voltage characteristics of extraction during the simultaneous formation of an electron and ion beam for two modes: a fixed voltage of a block accelerating ions of 1.5 kV (APSU 2, Fig. 2) and variation of the voltage of the electron acceleration block (APSU 1, Fig. 2) - Fig. 5 (curve 1) and the second mode of operation, when the electron acceleration voltage was recorded, and the ion acceleration voltage varied.

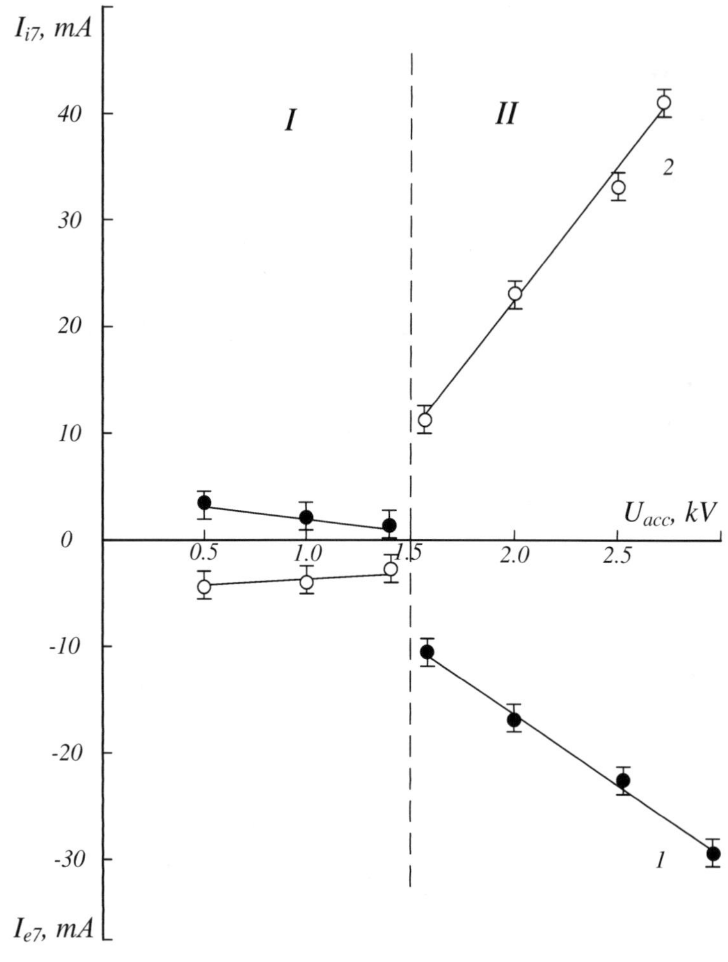

Fig. 5 - Current I_7 (into the Faraday cup). I_{e7} is the electron current in the Faraday cup. I_{i7} is the ionic current in the Faraday cup. (1) fixed ion acceleration voltage of 1.5 kV (APSU 2, Fig. 2). (2) fixed electron acceleration voltage of 1.5 kV (APSU 1, Fig. 3). The discharge current in chamber I (Fig. 1) is 200 mA, and the discharge voltage is 420 V. The discharge current in chamber II (Fig. 1) is 180 mA, and the discharge voltage is 410 V

In the case of a fixed ion acceleration voltage in section I (curve 1, Fig. 5), almost complete compensation of the electron beam is realized in the range from 0 to 1.5 kV and the current to the Faraday cup is close to zero. When the accelerating electrons voltage exceeds 1.5 kV (a fixed value of the ion acceleration voltage), the current to the Faraday cup rises, but it (region II on curve 1, Fig. 5) is lower than the emission current when the discharge is initiated in the discharge chambers I and II (Fig. 3, curve 2) and the supply of accelerating electrons to the voltage. When the electron acceleration voltage is fixed at 1.5 kV and the ion acceleration voltage is varied (region I, curve 2 in Fig. 5), the current into the Faraday cup undergoes a jump-like increase in the voltage region of 1.5 kV (Fig. 5, curve 2), which indicates the mutual compensation of the electron and ion beams up to this value and emission from the ion source at voltages above 1.5 kV. The value of the emission current at a voltage of 3 kV is 45 mA, and the density of the emission current is about 10 mA/cm^2, which indicates the prospects of using this design to develop a technological electron-ion source for implementing various processing technologies and surface modification of materials.

In the region of graphs I (Fig. 5), there is some current uncertainty, which is apparently caused by the presence of double electric layers in the discharge structures II and III (Fig. 1). In the area of graphs II (Fig. 5), i.e. at accelerating voltages of more than 1.5 kV, the field of double electric layers already has a weak effect on the movement of charges in gas-discharge structures II and III (Fig. 1), and the currents of electrons I_{e7} and I_{i7} ions (Fig. 5) have more definite values. It should be noted that the experiments were carried out for various gas inlets - in the range (1,1-3,5) mPa·m^3/sec. and discharge currents (0.18-0.24) A, the characteristics were similar to those presented in the figure not given.

The above characteristics generally indicate the possibility of using such a model for the development of a technological source of charged particles for the implementation of technologies that require combined exposure to electron and ion beams, and also serve as a prototype of a multi-discharge source cell to form an effect on large areas.

4. Conclusion

The presented designs of plasma sources of charged particles far from exhaust the entire spectrum of possible technological and structural solutions, but only show the potential capabilities of this type of sources for solving urgent problems of the formation of technological electron and ion beams. Such systems can form compensated ion beams, beams of neutral atoms or provide alternate or simultaneous exposure to beams of both types of charged particles, and, therefore, represent a unique universal tool for applying film coatings for various purposes [9, 11], which opens up new prospects for the implementation of more effective combined methods of exposure.

References

[1] Barchenko V T et al. 2014 *Physics and technology of plasma emission systems* (Saint Petersburg: SPbSETU «LETI»)

[2] Gruzdev V A, Zalesski V G, Antonovich D A, Golubev Y P. 2005 *Vacuum* 77 399-405.

[3] Zalesski V G, Antonovich D A. 2007 *Journal of Physics D: Applied Physics* **40** 7771-7777.

[4] Galansky V L et al. 1994 *Journal of Physics D: Applied Physics* **27** 953-961

[5] Bogomolov B K, Gilgenberg V V 2014 *12th International Conference on Actual Problems of Electronic Instrument Engineering, APEIE 2014* **1** 84-87.

[6] V Gruzdev V A, Zalesski V G 2010 *Plasma Physics Reports* **36**. 1191-1198

[7] Zalesski V G, Pobol I L, Gruzdev V A, Antonovich D A, Soldatenko P N 2016 *Herald of Polotsk State University. Series C, Fundamental sciences* **12** 37-44.

[8] Antonovich D A, Zalesski V G, Gruzdev V A, Soldatenko P N 2017 *Herald of Polotsk State University. Series C, Fundamental sciences* **4**, P. 45-51.

[9] Antonovich D A, Gruzdev V A, Zalesski V G, Pobol I L, Soldatenko P N 2017 *High Temperature Material Processes (An International Quarterly of High-Technology Plasma Processes)* **Vol. 21(2)**, p. 143-159

[10] F.M. Penning 1939 Coating by Cathode Disintegration. US Patent 2,146,025; N.V. Philips, Gloeilampenfabrieken, Eindhoven, The Netherlands.

[11] Bogomolov B K 2010 *10th International conference on actual problems of electronic instrument engineering proceedings. APEIE–2010* **1** 23-30.

Evaluation of influence of technological parameters on width of strengthened layer in plasma surface hardening of structural steels

A E Balanovsky[1], V E Gozbenko[2, 3], S K Kargapol'tsev[3], A I Karlina[1], Yu I Karlina[1]

[1] Irkutsk National Research Technical University, 83 Lermontov Street, Irkutsk, 664074, Russia
[2] Angarsk State Technical University, 60 Chaykovskoy Street, Angarsk, 668535, Russia
[3] Irkutsk State Transport University, 15 Chernyshevsky Street, Irkutsk, 664074, Russia
E-mail: fuco.64@mail.ru

Abstract. The article presents the results of the studies of the influence of technological parameters of plasma hardening on the geometric dimensions of the hardening track using modern research methods. The visualization of the heating spot on the metal surface according to the methodology made it possible to further evaluate the role of anode and cathode spots on the formation of a common heating spot in the process of plasma hardening of metals. It was established that, depending on the effective thermal power of the plasma arc, there are several modes of combustion of the plasma arc. For example, at an arc current of 20 A, there are several characteristic anode spots on the surface in the form of regular circles, which, with an increase in the arc current, contract to the center of the common heating spot. It has been established that the speed and distance of hardening affects the width of the hardened zone during plasma surface hardening of metals. The region of existence of several regimes of plasma hardening in a common heating spot was revealed, for which both single and group anode spots are characteristic.

1. Introduction

Due to the unique qualities of a compressed arc and a plasma jet as a source of heating the treated zone, plasma technologies affirmed in the general arsenal of other methods of surface treatment of metals [1-3]. A plasma jet and a compressed arc provide not only a thermal, but also a powerful force effect on the zone processing [4, 5]. The successful combination of high power and energy concentration in a compressed arc with the possibility of using a wide range of welding and auxiliary equipment designed for arc methods of metal processing suggests the prospect of using plasma technologies for surface hardening of various parts of machines and tools. At the same time, it must be recognized that very few papers have been devoted to theoretical issues of plasma surface hardening [1-3]. Moreover, the arsenal of plasma surface hardening technologies is expanding due to the development of new methods [4]. This paper presents the results of a study of plasma surface hardening from the standpoint of assessing the behavior and nature of the combustion of a plasma arc in argon. It is known [5-12] that the arc is a kind of gas insert between the electrodes and, like any conductor, interacts with magnetic fields. In this case, the pillar of the plasma arc can be considered as a flexible conductor, which, under the influence of a magnetic field, can move like any other conductor, deform and lengthen [5-8]. This leads to the deflection of the arc in the opposite direction to the greater tension. From a technological point of view, the possibility of controlling a plasma arc opens up great prospects for processing extended surfaces [3,4].

2. Experimental methodology and materials

All work was carried out in the laboratory of hardening and surfacing of materials at Irkutsk National Research State Technical University. This laboratory is equipped with a plasma hardening and surfacing installation. The design of this installation includes several basic elements that perform their function. In the installation, it is possible to highlight such key nodes as:

- the working body (as a working body used the power source of the plasma torch, which produces hardening and surfacing);
- high-speed camera (allows fixing and monitor all processes occurring in the process of arc burning);
- a camera (allows fixing in color the process of formation of cathode and anode spots, as well as observing the size of the hardening zone, the formation of the weld pool and the nature of arc burning) [3,4];
- an oscilloscope (allows measuring and plotting the voltage and strength of the welding current throughout the process),
- as a movement, a special mobile installation was used, providing movement of the product itself relative to the welding head;
- to increase the width of the hardening zone, a magnetic expansion scheme is applied (based on an alternating current inductor).

3. Research results

During the study, plasma hardening of steel 45 was performed. In order to determine the width of the hardening zone, the surface was first treated mechanically (using sandpaper), then the steel was etched with a 5 % solution of nitric acid in alcohol. The figures show photographs of hardened surfaces. Figure 1 shows photographs of a surface hardened at a speed of 11 mm / s with an arc gap of 3 mm after machining and chemical etching. Table 1 shows the width of the hardened zone of the plate.

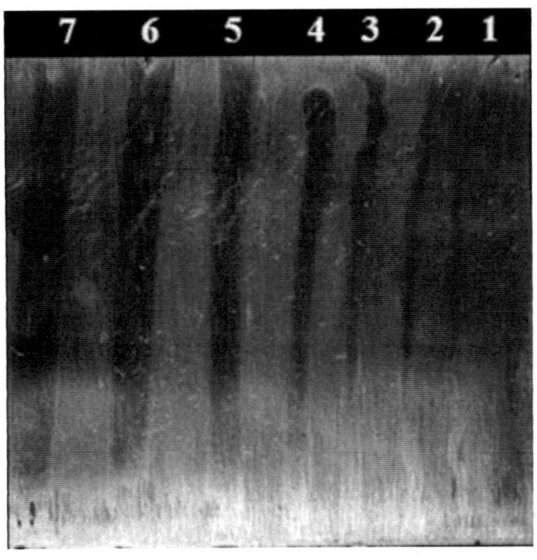

Figure 1. Hardening speed 11 mm / s, arc length 3 mm, surface after etching.

Table 1. Width of the zone of the hardened plate, mm.

Hardening area	1	2	3	4	5	6	7
	20A	30A	50A	70A	90A	110A	120A
Start	1.7	2.0	3.4	3.9	4.2	5.0	5.0
Average	1.0	1.5	1.8	2.7	3.8	4.6	4.8
Final	1.0	2.0	2.7	3.0	3.0	4.4	4.5

Figure 2 shows photographs of the surface hardened at a speed of 11 mm / s with an arc gap of 8 mm after hardening, as well as after mechanical processing and chemical etching. Table 2 shows the width of the hardened zone of the hardened plate.

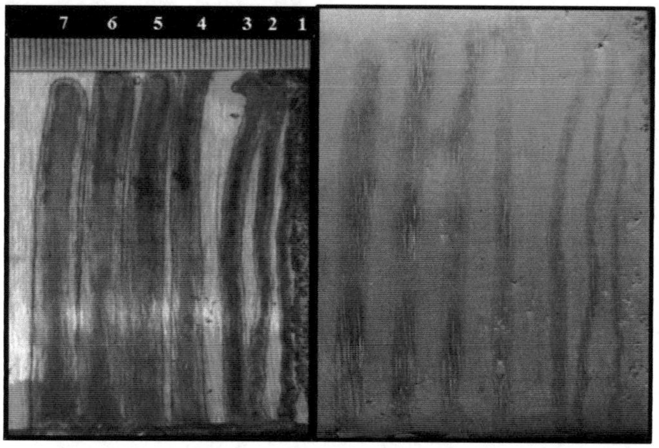

Figure 2. Hardening speed 11 mm / s arc length 8 mm, plate before and after etching.

Table 2. Width of the zone of the hardened plate, mm.

Hardening area	1	2	3	4	5	6	7
	20A	30A	50A	70A	90A	110A	120A
Start	1.4	1.9	2	2.5	4	4.5	5.5
Average	1	1.3	2.2	3	3.5	4	4.3
Final	1.3	1.5	2.2	2.8	3.7	3.8	4.5

Figure 3 shows the dependence of the width of the hardening track on the power of the plasma arc at different hardening distances (at a speed of 3.5 mm / s). It is seen that with increasing power and hardening distance, the width of the track increases. By analyzing the results obtained and comparing them with the results of [1-4], it is possible to draw a conclusion about the significant contribution to the geometric parameters of the hardening track of such parameters as processing speed and processing distance. Subsequently, the role of the cathode and anode spots of the plasma arc on the formation of the heating spot was evaluated according to the method described in [3,4,6-8].

Figures 4-6 show photographs taken at a welding current of 20-70 A. During hardening, it was noticed that the arc at a current of 20 A (Figure 5) burns unstably, constantly forming points with fusion, regardless of the movement of the detail, in this case, the arc lengthens, breaks, and forms a new binding, which does not give a high-quality hardened layer.

Figure 5 shows an arc at a current of 50 amperes. In the process of hardening, stable burning of the welding arc was observed on these parameters, no bindings with full melting were observed, the disadvantage with these parameters is small deviations. Figure 6 shows an arc at a current strength of

70 amperes. During hardening, stable burning of the welding arc was observed on these parameters, no bindings with full melting were observed, the hardening process took place without deviations. With a further increase in the current strength to 90 A, the plasma arc burns without deviations, ensuring the stability of hardening.

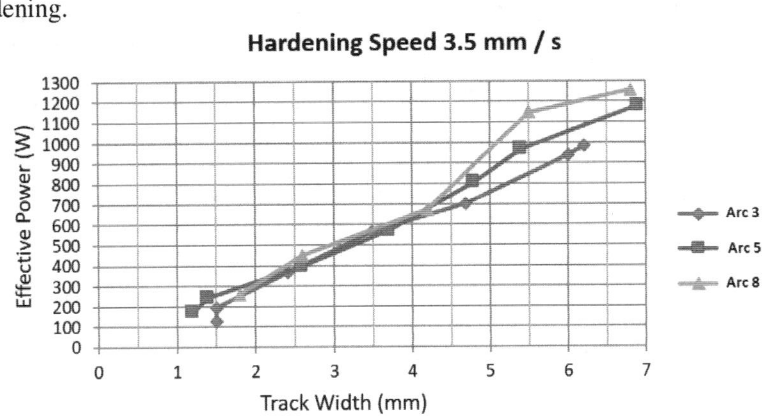

Figure 3. Dependence of the width of the hardening track on the power of the plasma arc at various hardening distances (at a speed of 3.5 mm / s).

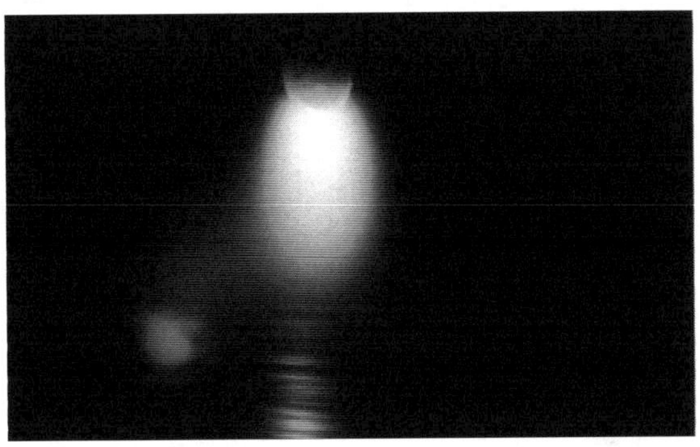

Figure 4. Photo of an arc snap made with force arc current 20 A.

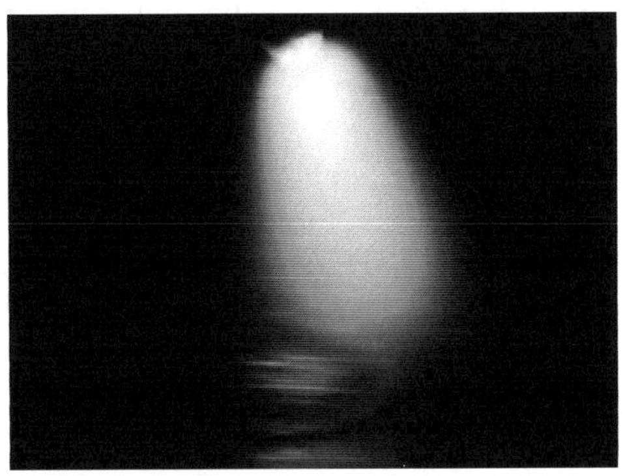

Figure 5. Photo of a heating spot made at an arc current of 50 A.

Figure 6. Photo of a heating spot performed at arc current 70 A.

The visualization of the heating spot on the metal surface according to the methodology [3,4] made it possible to further evaluate the role of anode and cathode spots on the formation of a common heating

IOP Conf. Series: Materials Science and Engineering **759** (2020) 012002 doi:10.1088/1757-899X/759/1/012002

spot in the process of plasma hardening of metals. It was established that, depending on the effective thermal power of the plasma arc [3,4], there are several modes of combustion of the plasma arc. For example, at an arc current of 20 A, there are several characteristic anode spots on the surface in the form of regular circles, which, with an increase in the arc current, contract to the center of the common heating spot. It was shown in [3,4] that the heating spot of the plasma arc has topologies that are nonuniform in temperature, which should affect the efficiency of heat input into the surface layers of the metal. In our experiments, we also fix the inhomogeneous distribution of heat along the width of the hardened track. It was shown in [5-12] that cathode spots can exert a thermal effect on a metal surface. In our experiments, cathode spots are grouped within the boundaries of the heating spot and most likely relate to spots of type 2 according to the classification of work [5,6]. It can be assumed that the type of spot will have a decisive effect on the formation of the surface layer structure and, therefore, on wear -resistance [4].

Thus, our studies have shown the importance of such technological parameters as the speed and distance of processing on the width of the hardened layer. Evaluation of the role of anode and cathode spots requires further continuation of work with the goal of revealing the general laws of the formation of a heating spot during plasma hardening.

4. Conclusion

In conclusion, the following conclusions should be noted:

1. It has been established that the speed and distance of hardening affects the width of the hardened zone during plasma surface hardening of metals.
2. The region of existence of several regimes of plasma hardening in a common heating spot was revealed, for which both single and group anode spots are characteristic.

References

[1] Leshchinsky L K, Samotugin S S, Pirch I I, Komar V I 1990 *Plasma surface hardening Publishing house* (K.: Tekhnika) p 109

[2] Korotkov V A, Berdnikov A A, Tolstov I A 1993 *Recovery and hardening of parts and tools by plasma technology* (Chelyabinsk: Metal) pp 144

[3] Balanovskii A E 2018 *High Temperature* **56(1)** 3-13

[4] Balanovskii A E, Vu Van Huy 2018 *Journal of Friction and Wear* Vol. 39 **4** pp 311-318

[5] Lyubimov G A, Rakhovsky V I 1978 *UFN* Vol. 125 **4** p 665

[6] Daalder J E 1976 *J. Phys* Vol. 9 p 2379

[7] Daalder J E 1979 *J. Phys* Vol. 12 p 1769

[8] Beilisy I, Djakovz B E, Juttner B, Pursch H 1997 *J. Phys* Vol. 30 p 119

[9] Schiilke T, 1996 *IEEE Transactions of plasma scien* Vol. 24 p 63

[10] Siemroth P, Schiilke T, Witke T 1997 *IEEE Transactions of plasma science* Vol. 25 p 571

[11] Sato A, Iwao T, Yumoto M 2008 *Plasma Sources Science and Technology* Vol. 17 pp 5

[12] Juttner B 1982 *Physica C* Vol. 114 p 225

Non-vacuum electron beam welding and cutting of cupper

T Hassel[1], A Beniyash[1,2], G Klimov [1]

[1] IW - Institute of Materials Science, Leibniz Universität Hannover, An der Universität 2, 30823, Garbsen, Germany
[2] email: beniyash@iw.uni-hannover.de

Abstract. Due to its physical properties, like high thermal and electrical conductivity copper is an indispensable material in many branches of industries. However the welding and cutting of copper and its alloys is challenging due to this properties and tendency to absorb gases resulting some defects like porosity or hot cracking. To prevent these, different requirements concerning the welding process have to be fulfilled. One requirement is the use of a highly concentrated heat input. The electron beam in atmosphere (NVEBW)is a suitable tool, which meets this requirement. This publication presents the advantages of NVEBW welding and cutting of copper. The experimental results of low-voltage (NVEBW) of copper will be shown and discussed too. The first portion studied the effects of varying power, travel speed and work distance on mechanical and electrical properties and structure of weldment at 175 kV accelerating voltage. The second portion studied the possibility of application of low-voltage 60kV NVEBW for welding of copper and influence of impurities in the copper on the welding parameters. The third portion studied the process of NVEB cutting of copper using the suction created by a local vacuum underneath the work piece. Experimental results of cutting with extremely high cutting speeds such as15m/min for 1.5 mm thick and 9m/min for 6mm thick copper will be representing.

1. Introduction

Copper materials are just as good weldable as steel materials taking into account the physical properties. However, a problem is the general tendency of non-ferrous metals to absorb atmospheric gases during welding; causing a reduction of the mechanical and technological properties of the weld seam. Therefore, all areas in which during welding occur the temperatures of more than 600 K must be protected from air access [1]. Other important properties for the welding of copper are its thermal conductivity as well as the thermal expansion. In comparison with alloyed steel, pure copper has an approx. 9 times higher thermal conductivity at room temperature (401 W/m°K for copper and 46 W/m°K 1132 for steel) and a 13 times higher thermal conductivity at 1000 °C (352W/m°K and 27 W/m°K) respectively. [2]. Compared to steel, thermal expansion increases by 1.4 times and approximately twice as much as shrinkage during solidification [1, 3], which causes a large welding deformation. The high thermal conductivity leads to dissipation of a large part of the introduced energy into the surrounding base material. The dissipated energy is no longer available for melting the base material. In the liquid state copper tends to absorb gases from the atmosphere, including oxygen and hydrogen. Oxygen in solid copper at the solidification temperature of the eutectic (1065 ° C) is only 0.09% soluble [4]. Already at low quantities is due to the eutectic reaction of the residual melt Cu formed oxide (Cu_2O). Hydrogen diffuses through the copper and reduced the inclusions of Cu_2O, forming H_2O (water), which then forms pressurized bubbles at the grain boundaries. This process can cause the grains to be forced away from each other, and is known as steam embrittlement or "hydrogen disease". [4] Based on the foregoing, should be selected welding technology.

Arc welding of copper and its alloys is widespread but challenging due physical properties of copper. To prevent the welding defects different requirements of the welding process have to be fulfilled. One requirement is the use of a highly concentrated heat input to minimise of weld pool size. Laser welding and electron beam welding (EBW) are the suitable technologies, which meets this requirement. However, these technologies are not without drawbacks.

Due to low absorption at room temperature, copper materials are classified as difficult to weld with state-of-the-art lasers Fig.1. The low absorption also causes high sensitivity to variations in surface conditions. Green laser radiation shows a considerable higher absorption at room temperature. So for copper SE-Cu58 at wavelength 515 nm absorption riches 42.8% [5]. However, green lasers have relatively low power and can only be used on thin material.

Electron Beam Welding in vacuum (EBW) technology is an established and widely adopted technique in industry. There are numerous applications of EBW in electrical, nuclear and automotive industry. EBW equipment covers a range of power from 2 to 200 kW. Vacuum chambers with volumes ranging from several litres (in the electronics industry) to several hundred cubic meters (aviation structures, heavy, petrochemical and nuclear) industries allow welding material from a fraction of a millimetre to 200 mm in thickness [6].

The main disadvantage of EBW in a vacuum is the presence of a vacuum chamber, which increases the cost of equipment and reduces productivity. The preparation of edges for EBW requires high precision. As a rule, welding is performed with maximum joint opening of 1% of the material thickness, which requires milling of the edges. During the welding of copper in vacuum intense evaporation takes place. This evaporation cause's formation of pores due to degassing of the molten metal in vacuum and to thinning of the weld seem. To reduce pore formation, it is necessary to increase the lifetime of the weld pool, which significantly reduces the welding speed. Reference [7] reports that defect-free welds of copper with thickness of 2-4 mm are obtained at speeds 0.3m / min or less.

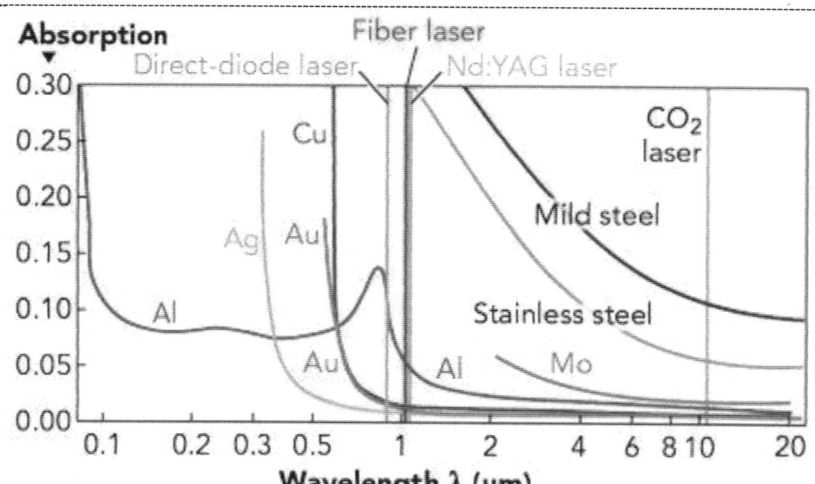

Figure 1. Light absorption by various metals, depending on the wavelength for normal incidence at room temperature [8].

Electron beam welding in the atmosphere (NV-EBW) is free of a significant part of the above disadvantages.

Non-vacuum electron beam welding employs essentially the same equipment as vacuum electron beam welding except that the working chamber is replaced by an orifice system [9, 10]. The electron beam emerges from the gun column via a series of differentially pumped vacuum stages, which are separated by small diameter orifices. Thus, the need for evacuation time is eliminated, as the orifice system and the generator column are permanently kept under vacuum. The very high welding speed and adaptability to automation makes NV-EBW especially useful for mass production with very high output [11]. Welding speeds up to 60 m/min for 1 mm thickness sheet are possible, as well as welding with deep penetration up to 20 mm at 0.5 m/min, which is significantly higher than laser or EBW in vacuum [12, 13], Compared to the laser beam or EBW, the electron beam on atmosphere has a more divergent profile. This guarantees a good gap bridging ability and the welding process has a wide tolerance towards inaccurate positioning. This technology is a promising alternative to laser and EBW in vacuum for welding and cutting of copper.

2. Experimental Setup

The welding tests were performed using a Non Vacuum Electron Beam Welding machine (PTR GmbH, Germany), located at the Institute of Materials Science at the University of Hanover, Fig. 2. The system has a maximum high voltage of 175 kV and beam power of 25 kW. For 3-D welding operations a CNC-work table with a travel length of 3.0 x 0.8 m in x-, y- direction can be used. A welding wire feeding system is fully integrated in the NVEB welder. Furthermore, the machine allows welding with pulsed beam current with variable frequencies between one and 999 Hz and pulse/break-ratios between 0 and 100 %.Principle set-up of NVEB is shown in Fig. 3.

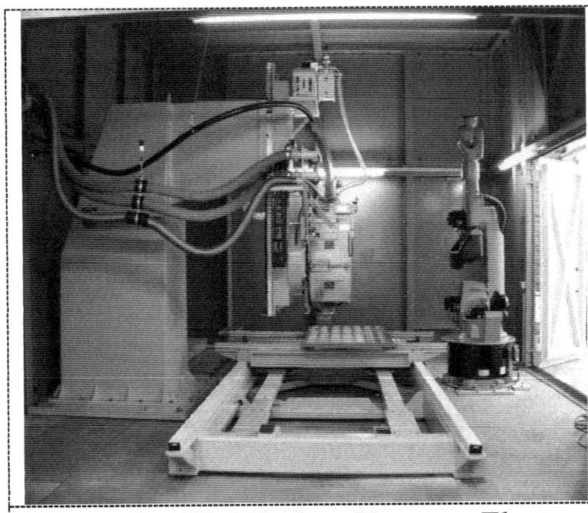

Figure 2 175 kV-Non Vacuum Electron Beam Welding machine in the Institute of Materials Science, University of Hanover.

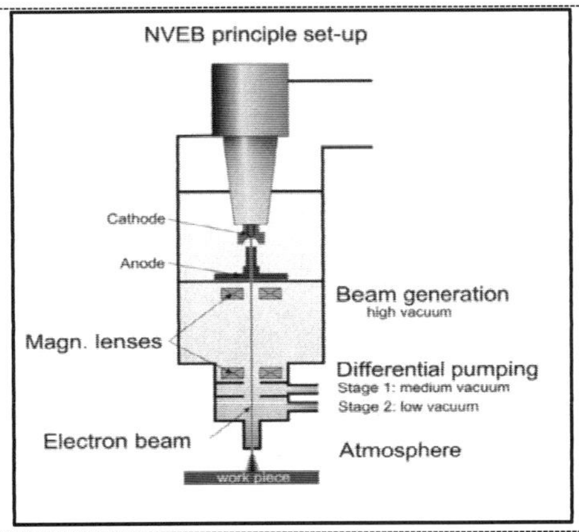

Figure 3 Principle set-up of NVEBW facility.

Welding tests were performed on sheets of cupper with thicknesses of 0.8; 1.5 und 6 mm. Physical properties and composition of copper samples are given in the Table 1.

Table 1. Composition and physical properties of various grades of pure copper

Cu-grade	Alloy components (mass fraction %)	Thermal conductivity W/m·°K	Specific electrical conductivity MS/m
Cu-OFE	–	393	58.0-59.1
Cu-HCP	P: 0.002-0.007	385	57.0-59.0
Cu-ETP	Bi: <0.0005 O: <0.0400 Pb: <0.0050	394	55-57
Cu-DHP	P: 0.0150-0.0400	305	43

3. Welding tests at 175 kV accelerating voltage

In experiments on a setup with an accelerating voltage of 175 kV, the influence of welding parameters on the properties of welded joints was studied. The high efficiency, the material-independent energy coupling in connection with a high total available power of 175kV machine make it possible to realize high welding speeds. The high thermal conductivity of copper leads to an increased dissipation of heat into the workpiece. Therefore, the relationship between thermal efficiency and welding parameters is very important. The thermal efficiency can be defined as being the ratio of the energy required to melt the weld bead to the total energy input over the same distance. The melting efficiency can be characterized used the linear energy, which representing the amount of energy per unit distance travelled. Figure 4 shows which linear energies have to be used to weld through 3 mm thick copper sheets using NVEBW. If one compares the necessary rated energy of 0.46 kJ / mm at 0.5 m / min, this can already be reduced to approximately one third (0.17 kJ / mm) at a welding speed of 4 m / min. Due to the high welding speed so less heat input into the workpiece can be achieved at the same penetration.

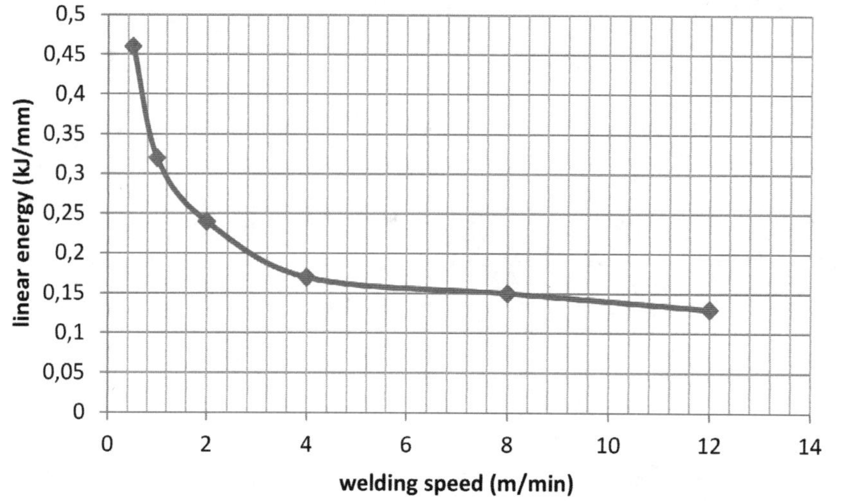

Figure 4. Linear energy vs welding speed for 3 mm DHP- copper

Despite the large power reserve of NV EB welder, an increase in welding speed has a limit. The welding speed is limited of such weld bead instability as humping or undercut bead. Humping can be defined as periodic undulations of the weld with undercutting along the edges of the weld seem Fig.5. Until now, there is no generally accepted theory of the origin of humping. The Rayleigh instability mechanism bears apparent similarities to the humping phenomenon, so it has been adopted and extended by many researchers to explain and describe the shape of molten beads deposited on a solid surface and also to determine conditions for instability by capillary forces [14, 15]. Other models are based on the flow of molten metal in a weld pool due to the Marangoni effect [16, 17]. Two main factors: the welding speed and penetration affect the shape of the weld. With the increase in the penetration depth, the threshold speed of undercuts and humping decreases. In our experiments with 3 mm thick, through-welded sheets at working distance of 15 mm with welding speeds up to 8 m/min no humping has been observed. At penetration depths in the range of 6 mm, the formation of humping could be observed at the welding speed of 3 m / min. In general, humping comes out of the weld pool dynamic, so factors that affect the dynamics may also have an effect on humping. This includes the working distance as a parameter, which influences the beam diameter and the power density via the scattering in the atmosphere.

In our experiments the copper sheets of 1.5mm, 3 mm, 6 mm and 10 mm were examined Fig 6,7. The welding parameters are given in the Table 2. Due to its low viscosity, copper melt tends to flow out of the weld. For the 3 mm sheets, a sagging of the weld root of 0.2 mm to 0.3 mm can be observed. For sheets with a thickness of more than 3 mm, a bath support must be used.

Table 2. Welding parameters for copper sheets at NVEBW with175 kV accelerating voltage

Material	Thickness (mm)	Beam current (mA)	Welding speed (m/min)	Work distance (mm)	Remarks
DHP	1.5	45	10	15	good
DHP	1.5	67	14	15	undercut
DHP	1.5	90	15	20	humping
DHP	3	80	8	15	good
ETP	6	103	5	8	good
ETP	10	140	2	10	good

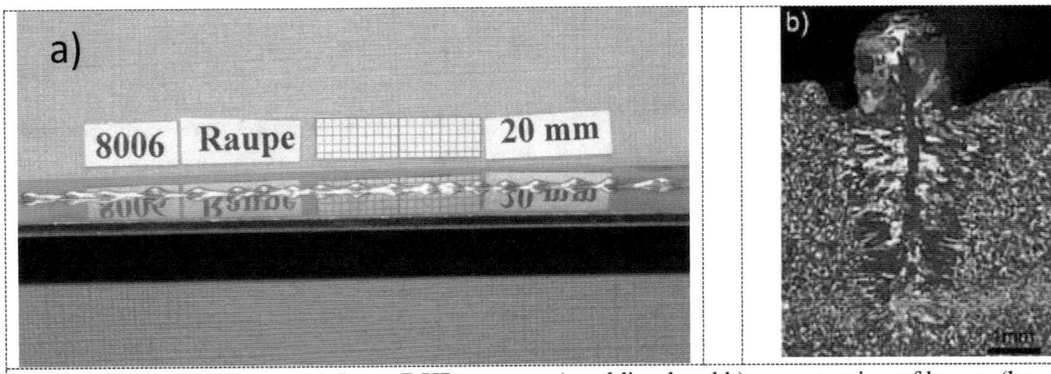

Figure 5. The Humping on 1.5 mm DHP-copper, a) welding bead b) cross-section of hump, (beam current 90 mA, work distance20 mm, welding speed 15m/min)

Figure 6. Weld seam and macro-section of of copper ETP 6 mm Plate (welding speed 5m/min, working distance 8 mm, beam current 103mA)

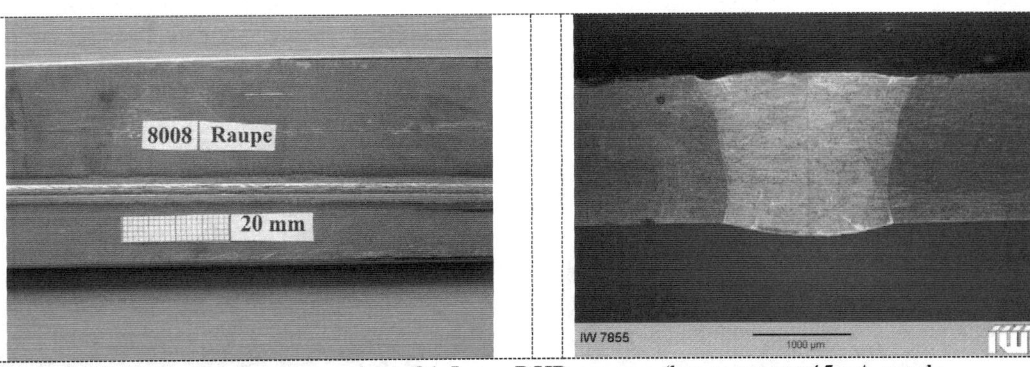

Figure 7. Weld seam and cross-section of 1.5 mm DHP-copper, (beam current45mA, work distance15 mm, welding speed 10m/min)

Figure 8. Microstructure of FZ Cu –ETP, 6mm, (mag.x50).

In the cross section of the welded joint Fig. 8, can be seen the fusion zone (FZ) consists of columnar crystal which crystal axes are directed normal to the diffusion line, and HAZ which grain size was slightly affected by the heat input applied during the welding. The Cu2O inclusions in the weld were not detected.

As mentioned above, oxygen has a negative effect on the properties of copper. In conventional non-vacuum electron beam welding, the welding process takes place under atmospheric conditions without protective gas coverage. Since oxygen has an influence on the properties of the joint, such as pore or crack formation, finally, the uptake of oxygen has been considered using the example of the oxygen-containing copper Cu-ETP. Compared with the base material with oxygen content of 89 ppm, an increase to about 225 ppm oxygen content in the weld metal was detected. Additional, experiments were carried out under atmosphere of argon. Wherein only a slight increase in the oxygen content in the weld to about 108 ppm oxygen was found.

4. Welding Experiments with acceleration voltage 60 kV

Among in electron beam welding experts, it is considered impractical to use an accelerating voltage of less than 150 kV for NVEBW, because due to the strong scattering of a low-energy beam in the atmosphere, the power density in the work area is not high. On the other hand, the requirements for the X-ray protection can be reduced and EB Generator can be small. For some applications such as brazing or welding of thin materials NVEBW at low acceleration voltage is of the great interest. Recently, in the Institute of Materials Science at the Leibniz University Hannover were carried out successful experiments on welding in the atmosphere at an accelerating voltage of 60 kV [18]. In the experiments, copper sheets of DHP and ETP grades with a thickness of 0.8, 1.5 and 2 mm were used. The maximum welding speed on sheets with a thickness of 0.8 mm reached 9m / min. The welding parameters are presented in Table III. Fig 9,10,11 shown cross-section welds obtained at a welding speed of 2,6 and 9 m/min and a working distance of 10,6 and 4 mm respectively. An inhomogeneous weld without full penetration was observed by welding of 2 mm sheets Fig 12. The weld width was wider at the area near the ending point than the starting point area since the material at the beginning of seam is cold; the beam energy is insufficient to melt the metal.

Figure 9. Cross section welds of Cu-ETP 1.5 mm, 2m/min, 60kV,130mA, 10mm

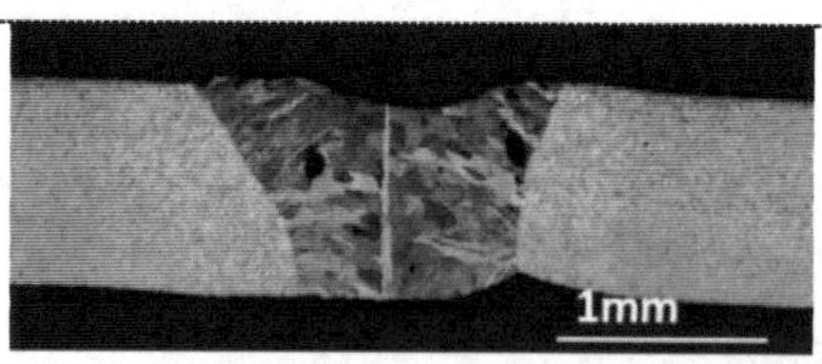

Figure 10. Cross section welds of Cu-DHP 0.8 mm, 6m/min, 60kV,120mA, 6mm

IOP Conf. Series: Materials Science and Engineering **759** (2020) 012003 doi:10.1088/1757-899X/759/1/012003

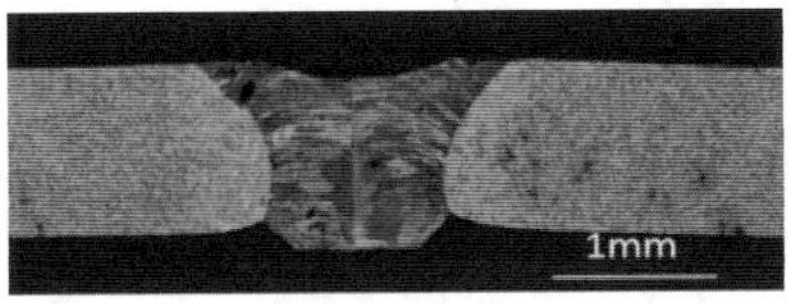

Figure 11. Cross section welds of Cu-DHP 0.8 mm, 9m/min, 60kV,120mA, 4mm

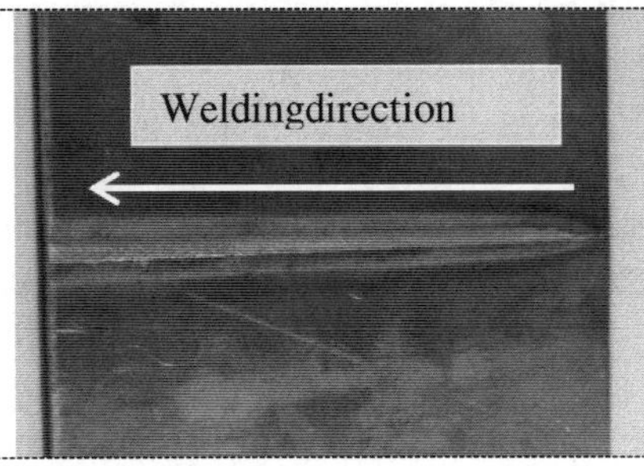

Figure12. An inhomogeneous weld of 2mm Cu-ETP.

Tensile strength testing of the welds of 0.8 mm ETP copper was conducted in compliance with the DIN EN ISO 6892-1 at room temperature using universal tensile testing machine Zwick Z100 operating with a crosshead speed of 0.5 mm/s. Dimensions of a tensile testing sample are illustrated in Fig.13 Results of tensile strength testing are summarized in Table 3.

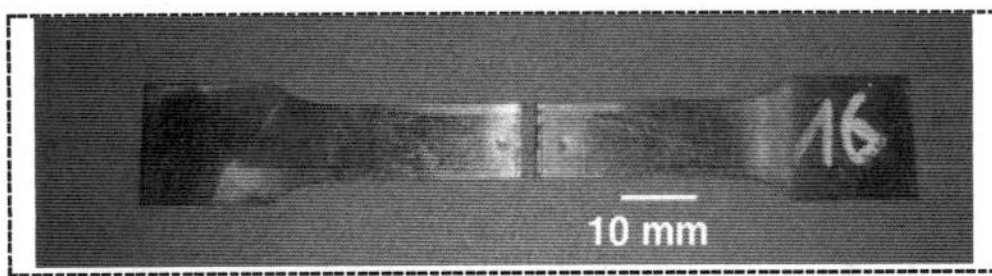

Figure 13. Copper sample 0.8 mm thick after tensile strength testing

The tensile strength of the weld reduced average 23% in comparison with the base material. The sample failed at the centre of the FZ, as indicated in Fig. 13. Fig. 14 shows the comparison of mechanical properties between welded samples and a reference sample.

Table 3. The tensile test of the weld samples

Nr.	$R_{p0.2}$ [MPa]	R_m [MPa]	Failure area
base	186.01	252.42	base
3	137.78	165.95	FZ
9	151.87	188.99	FZ
16	137.82	146.19	FZ

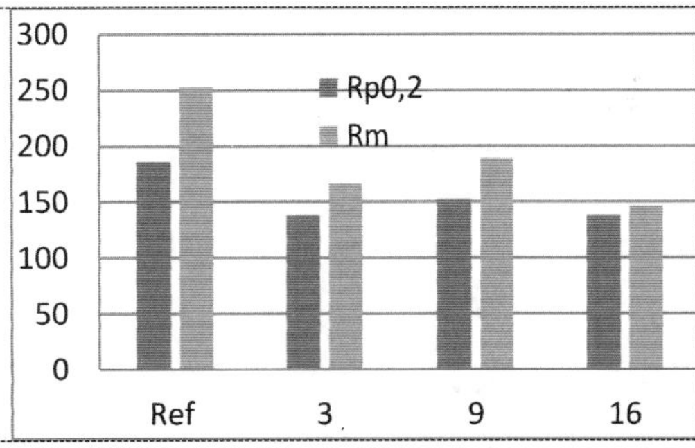

Figure 14. The comparison of mechanical properties between welded samples and a reference sample

5. Electrical conductivity of the weld

Figure 15 shows a diagram for measuring the electrical conductivity of welded samples. The voltage drop was measured at a distance of 100 mm and 10mm at the current of 100 amperes for 2 seconds are shown in the Table IV. In the weld area, there is a slight increase in voltage drop compared to the base metal. The electrical properties of pure cooper will only be affected when impurities are formed in the welds. The electrical measurement showed that welds was not significantly modified after the NV EB welding process.

Table 4. The voltage drop at the 100A current

Nr. spec.	voltage drop mV (10mm)	voltage drop mV (100mm)
3	2.15	18.11
9	2.06	18.07
16	2.16	18.09
Ref.	1.95	18.10

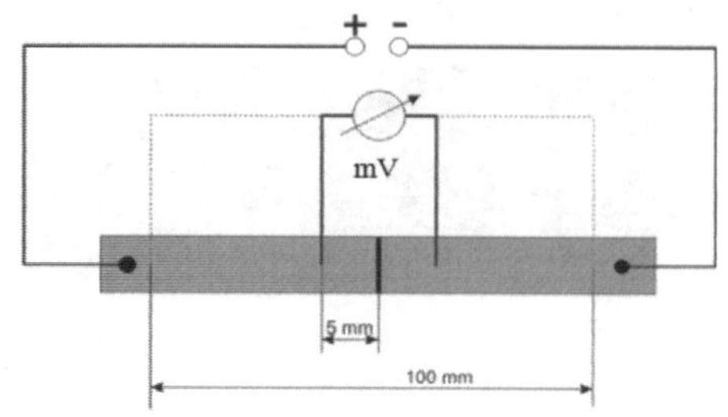

Figure 15. Principal set-up for measuring the electrical conductivity of welded samples.

6. Electron beam cutting

Most thermal cutting processes employ a concentrated gas jet directed onto the process zone to remove molten metal from the kerf. Contrasting to conventional cutting processes, we were able to realize a cutting process that utilizes a local low vacuum underneath the process zone. This is achieved by creating an area of low-vacuum between 100 and 500 mbar using a sliding seal. The electron beam melts the material and due to the pressure gradient between two sides of the workpiece induces a strong gas flow across the melting front, carrying molten and evaporated metal as well as smoke away from the kerf. This approach avoids high stagnation pressure, which will increase the beam power density and therefore will achieve very high cutting speeds. The gas flow can interact precisely at the melt front and will force the molten metal downwards. In another way than for a trailing gas jet, there is no dependence on the direction of the cut, so that curved contours can easily be realized. Experimental set-up for NVEB cutting with local suction were developed in Institute of Material Science of Leibniz University of Hannover. [19, 20]. It is important to remark that this method requires no change to the beam generator or pressure stage system of the original NVEB welder. Cutting experiments have been done with copper plates of 6 mm thickness. The 6 mm plates could be safely cut with a speed of 9 m/min at a working distance of 2 mm. The cut edges show a small ridge of residual melt droplets at the bottom edge. Macro- and micro-section examinations of these specimens show no sign of thermal influence induced by the cutting process Fig.16,17. The cut surface shows typical cutting groves for such materials, the JIS-B0601 (1994) roughness Rz is 135µm Fig.18 The width of the kerf is 1.4 mm on the top cut edge and 1.00 mm at the bottom cut edge measured from the macro-section profile. This agrees with an EN ISO 9013 quality grade of 3 for angular deviation.

Figure 16. Macro-section of NVEB-cut of 6 mm thick ETP-copper with 9 m/min at 24.5 kW.

Figure 17. Micro-section of NVEB-cut of 6 mm thick ETP-copper with 9 m/min at 24.5 kW.(mag.x50)

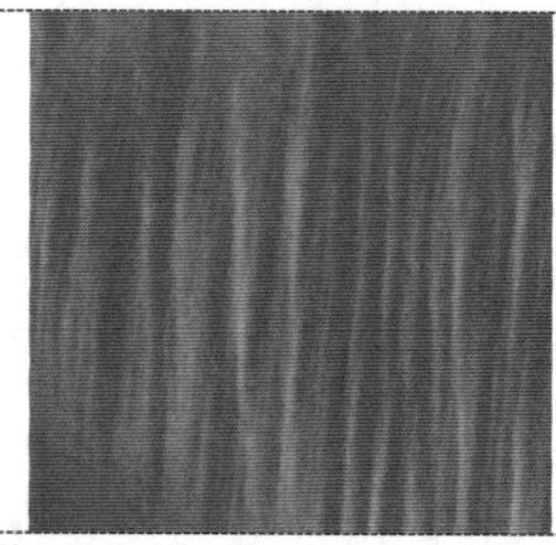

Figure 18. 3D laser scan of the cut edges

7. Conclusions

1. The non-vacuum electron beam (NVEB), due to its high power density, is a suitable tool for welding and cutting copper.
2. Due to the high welding speed and small size of the weld pool, the influence of the welding cycle on the material properties is minimal
3. Welding speed is limited by the appearance of humping
4. A low-energy beam with an accelerating voltage of 60 kV allows welding of copper sheets up to 2 mm thick with high speed.
5. No significant changes of electrical resistivity and chemical composition of welds were observed after welding.
6. A new cutting method with local low vacuum underneath the process zone allows cutting of 6mm copper at speeds up to 9 m / min with high quality cuts.

References

[1] Schulze G 2004 Die Metallurgie des Schweißens 3 Auflage (Berlin: Springer-Verlag).
[2] Handbook of chemistry and physics 2007 88th. edition (CRC Press).
[3] Dorn L 1987 Metallphysikalische Vorgange beim Schweißen von Kupfer und Kupferlegierungen– Werkstoffliche, *Grundlagen. Schweißen und Schneiden* **Nr.12 39**.
[4] Schweißen von Kupfer und Kupferlegierungen, Deutsche Kupferinstitut Informationsdrück i.12 Auflage 2009.
[5] Englera S, Ramsayera R, Poprawe R 2011 Process studies on laser welding of copper with brilliant green and infrared lasers *Physics Procedia* **12** 339–346.
[6] Dobeneck D 2007 An international history of electron beam welding, 1st edition (Germany: Pro-beam AG).
[7] Nazarenko O et.al. 1987 Electron beam welding [in Russ.] (Kiev: Naukova Dumka)
[8] AKELA LC 2019 Laser diode cutting. URL: https://www.akelalaser.com/markets/industrial/diode-laser-cutting/ .
[9] Arata Y, Tomie M 1970 Some fundamental properties of non-vacuum electron beam *JWS* **September**.
[10] Powers D E 1997 Nonvacuum electron beam welding enhances automotive manufacturing *Welding Journal* **76**.
[11] Hinse-Stern A, Schwab U 2004 Elektronenstrahlschweißen an Atmosphäre (NVEBW) von Modulträgern 6 International Konferenz Strahltechnik 26.04–28.04.2004.
[12] Draugelates U et.al 2000 Hochgeschwindigkeit Elektronenstrahlschweißen von Aluminium unter Atmosphärendrück *Schweißen und Schneiden* **52**.
[13] Bach Fr.-W, Beniyash A, Lau K, Konya R 2009 Non-vacuum electron beam welding of structural steels *Paton Welding Journal* **May**.
[14] Arata Y Tomie M 1970 Some fundamental properties of non-vacuum electron beam *JWS* **September**.
[15] Bradstreet B J 1968 Effect of surface tension and metal flow on weld bead formation *Weld J* **47** (7) 314–22.
[16] Mills K G, Keene B J 1990 Factor affecting variable welding penetration *International Material Review* **35** 185-216.
[17] Reisgen U et al 2012 A Investigation of factors influencing the formation of weld defects in non-vacuum electron beam welding *The Paton Welding Journal* **2** 11-18.
[18] Hassel T, Klimov I, Beniyash A 2018 Beam extraction using non vacuum electron beam by reduced acceleration voltage beam technologies and laser application *Journal of Physics Conf Series* **1109**.

[19] Hassel T, Murray N, Klimov G, Beniyash A 2016 Cutting and welding of high-strength steels using non-vacuum electron beam as a universal tool for material processing *World Journal of Engineering and Technology* **4** 598-607.

[20] Beniyash A, Murray N, Klimov G, Hassel T 2017 Using a non-vacuum electron beam as a universal tool for material processing In Proc. 2nd International conference on Electron Beam Wielding (14.11.2017-17.11.2017) (Moscow: MPEI).

IOP Conf. Series: Materials Science and Engineering **759** (2020) 012004 doi:10.1088/1757-899X/759/1/012004

Effects of laser treatment modes on the parameters of the hardening zones and the wear resistance of steel 40Cr

V P Biryukov [1], V N Petrovsky [2], D V Panov [2], D V Ushakov [2]
[1] Mechanical Engineering Research Institute named by A.A. Blagonravov, 4, Maly Kharitonyevsky Pereulok, Moscow, 101990, Russia
[2] National Research Nuclear University MEPhI, 31, Kashirskoye highway, Moscow, 115409, Russia

E-mail: laser-52@yandex.ru

Abstract. The paper presents the results of metallographic and tribological studies of laser hardening zones of steel samples. A full factorial experiment on laser hardening with a change in the distance from the focal plane, radiation power and processing speed was carried out. The regularities of changes in the depth and width of the quenching zones are obtained from the regression equations. With low-frequency transverse scanning of the laser beam up to 100 Hz, the hardening zones with a width of 20 mm and a depth of 1.4-2.0 mm in one pass were obtained. The wear resistance of the hardened zones is 3.68 times higher than the base material.

1. Introduction

Laser hardening of martensitic stainless steel AISI 416 with a carbon content of 0.167% was carried out on samples with dimensions of 55×10×7.5 mm with an initial hardness of 155 HV [1]. The Nd:YAG laser of Rofin-Sinar company with a maximum radiation power of 2.2 kW was used. The laser beam was focused to a diameter of 2 mm on the sample surface. The testing regimes of laser hardening was carried out at radiation power of 0.7 and 1 kW when the velocity of the beam of 0.5; 1; 2; and 3 m/min. The maximum depth of hardening of 0.9 mm with a hardness of 400-700 HV obtained with a radiation power of 1 kW and a speed of displacement of the beam 0.5 m/min. the wear Tests were performed according to the scheme of the disk (diameter 73 mm, hardness 63 HRC) – pin (7×7,5×10 mm). The linear velocity of the disk 8.4 m / min., the Load on the image under test 50 N. Found that the minimum wear rate amounted to 0.001 g / min for the samples treated with the speed of movement of the beam of 0.5–1 m/min and output power of 0.7–1 kW. This corresponds to the wear intensity of the reference samples of this steel after volumetric hardening.

Laser heat treatment of the Ck45 Steel surface creates a microstructure with 91.65% needle martensite and 8.35% residual austenite [2]. Martensite hardness reaches up to 850 HV, and residual austenite 400-600 HV. Wear tests were performed according to the disk – finger scheme at a load of 30 N, the speed of movement of 0.7 - 1.49 cm/s. the Hardness of the disk was 385 HV. The duration of each test was 10 minutes. The wear resistance of laser-hardened samples is twice as high as that of the original steel.

The effect of the fiber laser beam defocusing on the depth and width of the laser hardening zones, processing modes on the parameters of the hardened zones at low-frequency transverse vibrations of the beam is determined.

IOP Conf. Series: Materials Science and Engineering **759** (2020) 012004 doi:10.1088/1757-899X/759/1/012004

2. Materials and methods

For laser hardening, 40Cr steel samples with dimensions of 12×20×70 mm were performed using a laser complex based on a fiber laser LS2 equipped with an optical head FLWD50L fixed on the movable flange of the KUKA robot arm. The diameter of the transport fiber 100 μm, focus collimating lens of 160 mm, the focusing lens of 500 mm. the Processing is performed with the radiation power of the laser 1000 and 1800 W, the speed of the beam 10 to 18 mm/s, the defocusing of the beam in the range of 25 – 200 mm.

Metallographic examination was performed with use of hardness testing of the PMT-3 under a load of 0.98 N, a digital microscope AM413ML. metallographic microscope Altami MET 1C.

In the second series of experiments using the method of full factorial experiment (PFE) the influence of processing modes on the parameters of hardened tracks was determined. The power of radiation P, W, processing speed V, mm/s, and defocusing of the beam Z, mm were chosen as experimental factors the depth H and width B of the laser hardening zones were considered as the responses of the system For the construction of mathematical models. Table 1 presents the initial data. In this paper, in addition to taking into account the influence of experimental factors on the response, all possible interactions of factors with each other were taken into account. The regression equation has the form [3]:

$$y = b_0 + b_1x_1 + b_2x_2 + b_3x_3 + b_{1,2}x_1x_2 + b_{1,3}x_1x_3 + b_{2,3}x_2x_3 + b_{1,2,3}x_1x_2x_3 \quad (1)$$

where:

y – system response;

x - encoded variables

b – coefficients of the regression equation.

Table 1. Summary table of coding of FFE variables

Factor x_i	Upper level factor x_i^+	Lower level factor x_i^-	Centre plan x_i^0	Range of variation Δ_i
P (W)	1800	1000	1400	400
V (mm/s)	18	10	14	4
Z (mm)	150	100	125	25

At the end of the experiments, the sections were made according to the standard procedure and triple measurements of the depth and width of the hardened zones were made. All possible interactions of factors were determined in the calculation. Since FFE 23 was performed, the number of experiments was 8 for each series.

In the third series of experiments, broadband laser quenching was performed at transverse oscillations of the beam relative to the velocity vector of the beam using a galvanoscanner with a mirror oscillation frequency of up to 100 Hz. Wear resistance tests were performed on the friction machine MTU-01 according to the scheme plane (sample) - ring (counter-tile, steel 40X, HRC 48-52) at a load of 2 MPa. As the lubricant used oil TP 22S.

3. Research results and discussion

According to the results of metallographic studies of hardened zones in the first series of experiments, graphs of the depth and width of the quenching zones on the change in the distance to the focal plane Z at a linear beam energy of 100 j/mm are presented in **Figure** 1 (a, b). The depth and width of the quenching zone varies almost linearly in the range of 100-150 mm and therefore this area can be described by first-order regression equations [4].

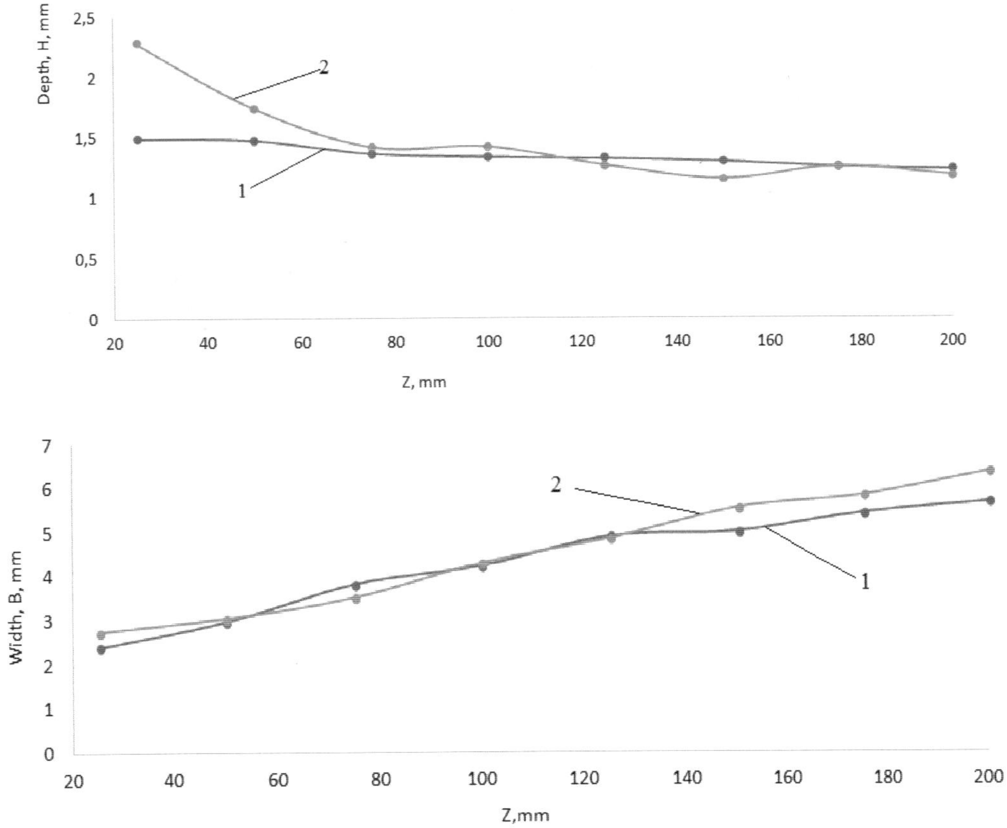

Figure 1. Dependence of the depth (top) and width (bottom) of the laser hardening zone of steel 40Cr from the beam defocus: 1 – V = 10 mm/s, P = 1000 W, 2 - V = 18 mm/s, P = 1800 W.

In the second series of experiments the regression equations in the defocusing of a beam of 100 – 150 mm. Depth of the hardening zone:

$$H = 1{,}25 + 0{,}196\frac{P-1400}{400} - 0{,}172\frac{V-14}{4} + 0{,}05\frac{Z-125}{25} - 0{,}001\frac{P-1400}{400}\frac{V-14}{4} - 0{,}043\frac{V-14}{4}\frac{Z-125}{25} - 0{,}049\frac{P-1400}{400}\frac{Z-125}{25} + 0{,}0072\frac{P-1400}{400}\frac{V-14}{4}\frac{Z-125}{25} \quad (2)$$

where
P – laser radiation power, W;
V – beam speed, mm/s;
Z – distance to the focal plane, mm.
The width of the hardening zone:

$$B = 4{,}701 + 0{,}37\frac{P-1400}{400} - 0{,}275\frac{V-14}{4} - 0{,}514\frac{Z-125}{25} + 0{,}102\frac{P-1400}{400}\frac{V-14}{4} + 0{,}134\frac{V-14}{4}\frac{Z-125}{25} + 0{,}250\frac{P-1400}{400}\frac{Z-125}{25} - 0{,}006\frac{P-1400}{400}\frac{V-14}{4}\frac{Z-125}{25} \quad (3)$$

The regression equations were calculated and the results were compared with experimental data. The calculated values differ from the actual depth and width of the quenching zones by no more than 5%. Regression models of dependencies of type H (P,V), B (P,V) are introduced into the MSExcel table editor and comparative surfaces for these functions are constructed (**figure** 2). The prevailing influence on the geometric parameters of the quenching zones is the radiation power (figure 2). With increasing power, the width and depth of the quenching zone grow. As the travel speed increases, the depth and width of the hardened zones decreases. However, at Z=100 mm at lower speeds, the width varies slightly (figure 2 b), which is due to the energy consumption for melting the surface of the sample material. As the beam defocus (diameter) increases, the depth of the quenching zones decreases and the width increases.

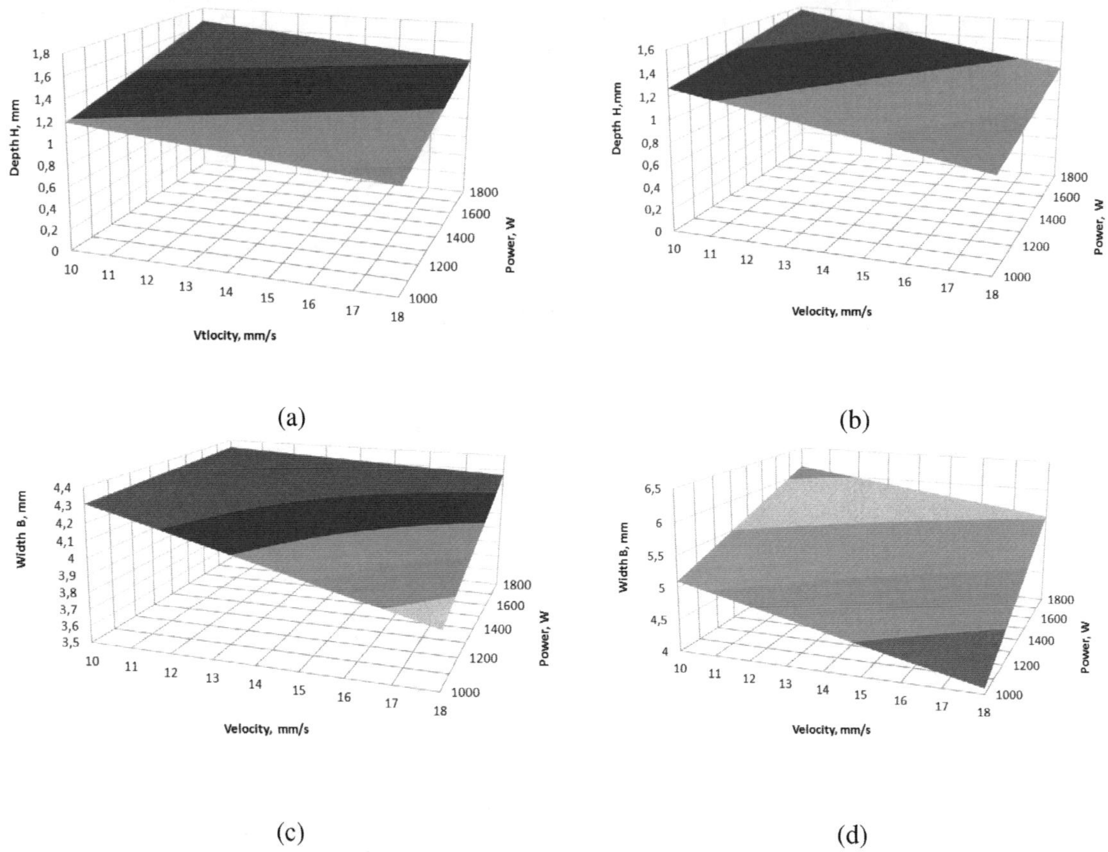

Figure 2. Influence of processing modes on the depth N and the width of the quenching zone at defocusing of the beam Z: and Z =100 mm and big Z =150 mm.

Figure 3 shows the micro-sections of the hardened steel 40X with the width and thickness of the hardening zone of 20 mm and 1.4-2.0 mm, respectively, obtained when the defocusing of the beam Z=150 mm, radiation power P=1800 W at various speeds of processing with the use of transverse vibrations of a beam up to 100 Hz. The microhardness of the laser hardening zones varied within 7240 – 7900 MPa.

The results of comparative wear tests are shown in figure 4. The first column shows the wear intensity J = $1.98 \cdot 10^{-9}$ of the substrate normalized steel 40KH. The second column of the wear intensity J = $0{,}532 \cdot 10^{-9}$ quenched samples with the width of the zone of 20 mm

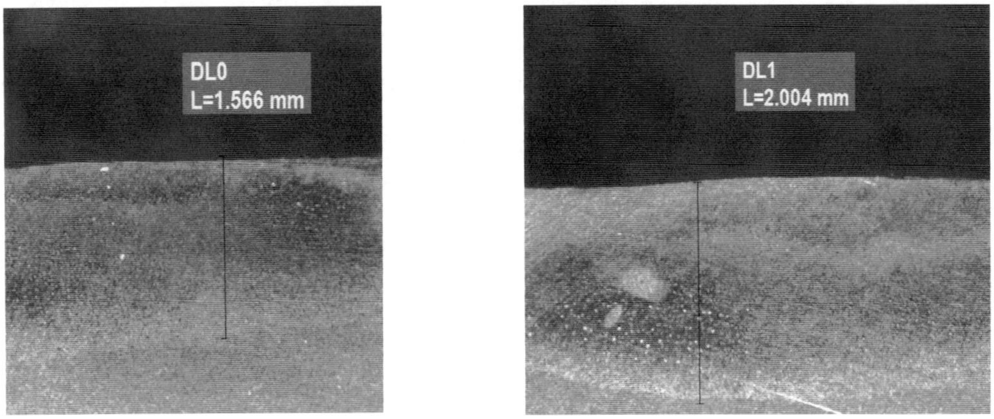

(a) (б)

Figure 3. The micro-sections of the zones of hardening of steel 40KH with a transverse beam oscillations up to 100 Hz

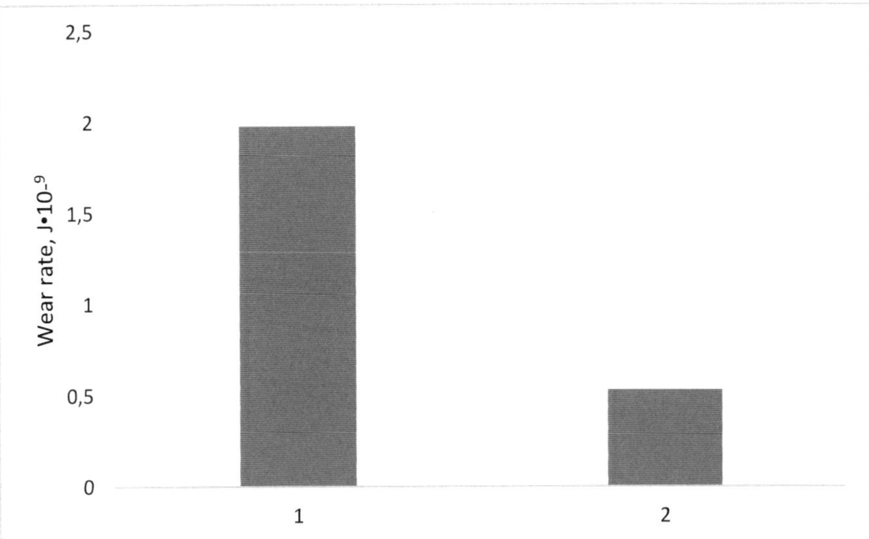

Figure 4. Wear test results: 1 – base material, 2 – hardened layer

4. Summary

Linear regression equations for beam defocusing of 100-150 mm are obtained to calculate the depth and width of the quenching zones with an error of not more than 4%. The constructed surfaces show the regularity of changing the depth and width of the hardened zones from the processing modes. The technology of laser hardening with a width of 20 mm with a layer depth of 1.5-2.0 mm in one pass at transverse vibrations of the beam up to 100 Hz has been worked out. The wear resistance of the friction surfaces with a broadband laser hardening is increased 3.68 times compared with the normalized steel 40KH.

References

[1] Al-Sayed S R 2017 Characterization of a Laser Surface-Treated Martensitic Stainless Steel *Materials* **10**

[2] Adel K M 2014 Enhancement of Dry Sliding Wear Characteristics of CK45 Steel Alloy by Laser Surface Hardening *Procedia Materials Science* **6** 1639 – 1643
[3] Evdokimov Yu A, Kolesnikov V I, Teterin A I 1980 *Planning and analysis of experiments in solving friction and wear problems.* Moscow. Science.
[4] Biryukov V P 2017 Experimental determination of parameters of hardened zones during laser hardening of cast irons and steels *Journal Photonics* **2** 22-32

Beam diagnostics on the electron beam welding installation at Budker Institute of Nuclear Physics

K A Blokhina[1,3], T A Devyataykina[1,2], A M Medvedev[1,2], Yu I Semenov[1], M M Sizov[1,2], A A Starostenko[1,2], A S Tsyganov[1]

[1] Budker Institute of Nuclear Physics, 11, Academician Lavrentiev ave., Novosibirsk, 630090, Russia
[2] Novosibirsk State University, 2, Pirogov ave., Novosibirsk, 630090, Russia
[3] E-mail: kseniyablokhina@yandex.ru

Abstract. This paper describes the equipment and methods of the beam diagnostics, namely, the beam current density distribution measuring and the beam position and inclines angle determination. Such a diagnosis is aimed at applying on the electron-beam technological installations, in particular, on the electron-beam welding installations. The paper presents the results of the beam current density distribution measurement, describes the design of the beam position and inclines angle measurement system, and presents the results of preliminary tests of the prototype of this system.

1. Introduction

Nowadays an electron beam is widely used tool both for scientific researches and for different applications. With the development of electron beam technologies the beam quality requirements are increasingly rigorous. Therefore, simultaneously with development equipment for beam generation and transportation the direction of beam quality diagnostics have to be developed. It is the fine-tuning of the beam that ensures the high quality of the manufactured products, and it also affects the repeatability of scientific experiments not directly related to electron beam processing but using electron beam.

One of the practically important electron beam implementations is electron beam welding. To obtain high quality weld seams it is necessary to monitor various electron beam parameters, first of all, the power density distribution in the beam cross-section.

The present work was carried out on EBW installation in Budker Institute of Nuclear Physics SB RAS, Novosibirsk [1]. The installation includes:

- 60 keV electron gun with current up to 750 mA;
- magnetic optical system with ability of deflection and parallel beam shift;
- two-coordinate sample holder with the ability to move;
- seam tracking system based on reflected electron sensing technique.

A significant part of the equipment was designed and manufactured directly at BINP. The work on developing new modules and subsystems is being carried out continuously.

This paper presents the results of the electron beam diagnostic tools development based on the external current collector signal processing.

2. Problem statement

To exam BINP electron guns beam quality the slit and wired detectors are being used [2, 3]. The measuring procedure is based on the following operating principle: a circular beam oscillation is applied to the gun deflecting systems, obtained signal from the detector is observed by an oscilloscope. Measured beam profiles allow estimating beam radial size (the result was 0.6 mm under 2.7 mA current).

Due to increasing the number of works carried out on BINP EBW installation, which require precise beam positioning, a new diagnostics tool realization able measured not only profile but beam power distribution was required.

3. Diagnostic technique choice substantiation

The simplest and widely used types of devices for measuring beam current and determining its profile is contact one. When using this device type, the beam profile is determined from the known speed of movement of the device relative to the beam, the device geometry and the measured time of exposure of the beam to a current collector.

In our case from mentioned type of devices the Faraday cup is considered to be the most suitable collector due to its high current measurement accuracy. While the task is not to measure the total beam current, different masks or diaphragms are placed before the collector. Masks are aimed at highlighting a certain part of the beam. Then the current distribution in the beam cross-section is determined.

As a supporting beam shape diagnostic a luminophore screen visualization could be used. Interacting with luminophore surface electrons lose part of their energy by ionization, in turn, part of the ionization loss is converted into optical radiation. The intensity of visible light from the luminophore surface is proportional to the beam current. Luminophore can be used for subsequent diagnostics with an optical sensor. At the moment of the optical beam diagnostics experiments have not been carried out yet but are planned in the future.

Therefore, the system for current density distribution measuring with a Faraday cup current collector is the only developed beam diagnostic system. The decision on the initial device design was based on the fact that it was planned to carry out beam diagnostics at the small pretuning beam current. Due to the low received power, the small dimensions of the device allow it being in the vacuum chamber constantly and do not impede welding process.

Graphite was considered as the most suitable mask material. Graphite has a high melting point, which allows it being exposed to the beam for a prolonged period. In addition it has a low electron reflection coefficient (~0), resulting in inappreciable signal distortion (for example, copper masks gave a poor signal to noise ratio). And the last important advantage of graphite is that it can be easily processed into complex shape, and this allows producing a replacement item quickly in case of damage.

As a result, a graphite plate with a 0.1 mm diameter round transcentric symmetrical pin-hole was chosen as a mask for the power density distribution system.

4. Beam current distribution measurement

4.1. Measurement procedure

4.1.1. Beam oscillation forming. The beam oscillation in the diaphragm plane is set by periodic changing of magnetic system fields that allows isolating and measuring the beam small part current passing through the diaphragm (See Figure 1).

ADC connected to the shunt R of the Faraday cup controls the beam position relative to the pin-hole (by ADC readingU_r). In the seam tracking system the same ADC data are used to measure the secondary emission electrode voltage drop and to observe the surface map [4]. The collected data arrays during scanning are written to a separate text file. Assuming the resistance R is constant, we find the current at each point $U_r(x) = I(x)R$.

Analytically, the beam current is calculated by the formula:

$$I = \int_{-\infty}^{\infty} J(x, y)dy,$$

where I(x) is the instantaneous cylinder current, and J(x, y) is the distribution of the current density in the beam per unit surface area.

Then, the power distribution in the beam is calculated by the formula:

$$P(x, y) = J(x, y)U_{ac},$$

assuming the accelerating voltage is constant.

Beam diagnostics are carried out with a beam current of ~ 1 mA, since a larger current can destroy the diagnostic system.

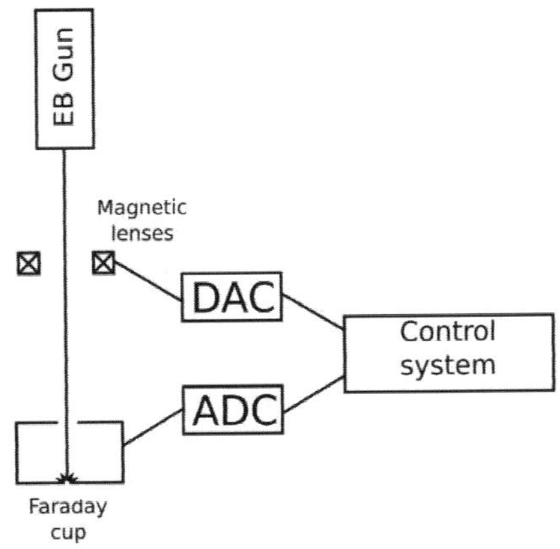

Figure 1. Measurement procedure scheme

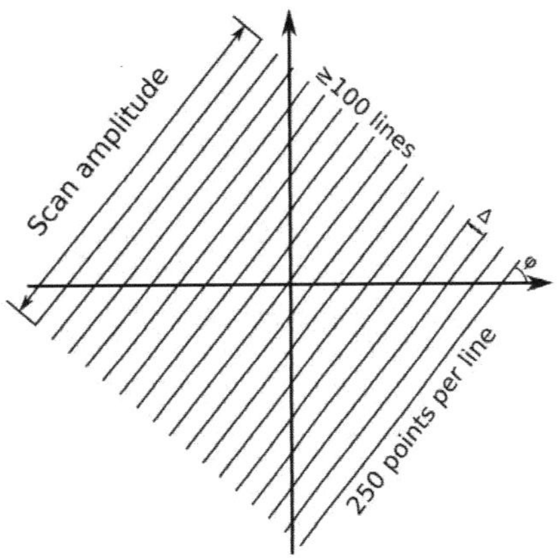

Figure 2. Scanning beam oscillation

4.1.2. Scanning function. The current values in a pair of magnetic correctors are set through the DAC. The values written to the DAC are set by a periodic function with a smooth current change (up to 500 points) in one direction and sequential bias in the other. The function is implemented in the Elixir functional language.

In each point the voltage drop is amplified and read by ADC. The scan system generates a symmetrical saw tooth signal with the step (Δ). The step value has to be set by operator (see Figure 2). Also operator can set such parameters as the angle of rotation (φ) and the scanning beam oscillation amplitude. For data processing and displaying Python Spyder development environment is used.

4.2. Beam current distribution composition algorithm

Scanning is performed on a raster of 250 points per line and ~ 100 lines per projection. After scanning in one projection, the scanning beam oscillation is rotated by a certain angle φ in the range from 0 to 90 degrees (see Fig. 2). For detailed current density distribution performing all the specified projections is required. Usually their number is from 7 to 15.

The noise in the signal of a single scan is too large; therefore, a parameter that controls the number of passes along one scan line was added to the scanning function. Collecting statistics significantly reduces the noise. The beam energy distribution is assumed to be constant over time. However, in order to avoid possible errors it is also necessary to collect statistics for averaging fluctuations in case of energy jump. To obtain reliable data the number of iterations in the existing system must be at least 10. Collecting data for a full beam scan with ≥100 scan lines takes about 1 min per projection. Figure 3a shows the result of sharply focused beam current distribution measurement. Figure 3 (b, c) shows two projections of current distribution for an insufficiently focused beam, where the irregular shape of the distribution can be observed in detail.

IOP Conf. Series: Materials Science and Engineering **759** (2020) 012005 doi:10.1088/1757-899X/759/1/012005

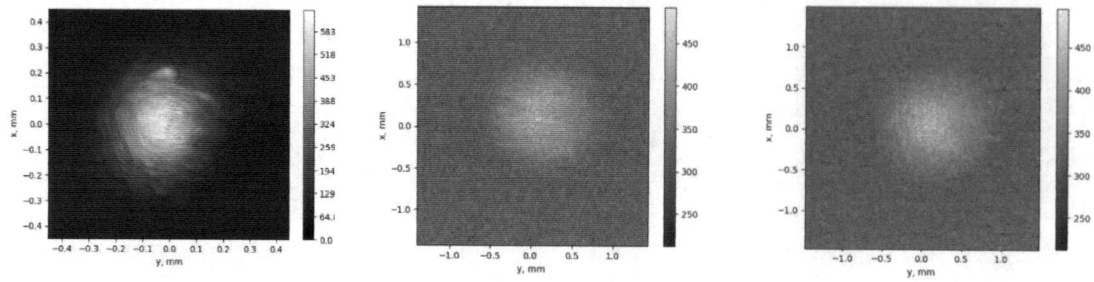

(a) (b) (c)

Figure 3. Beam current distribution (ADC readings per mm^2) with total current of 1.2 mA: (a) – beam current distribution at the sharp focusing; (b) – the x axis projection of the under focused beam; (c) - the y axis projection of the under focused beam.

4.3. Magnetic optics parameters selection algorithm

To select the required magnetic optics elements current values it is necessary to obtain current density distribution in one projection for each of the current values of the focusing lens or corrector. By variating the lens current the optimal focus depending on correspondent counted beam size is set. By variating the magnetic corrector current the anode correction is set according to the change in the waveform and the beam displacement nonlinearity. An example of the magnetic corrector values selection is shown on Figure 4 whence it is obvious that the y-corrector current value is set correctly and the x-corrector current must be increased.

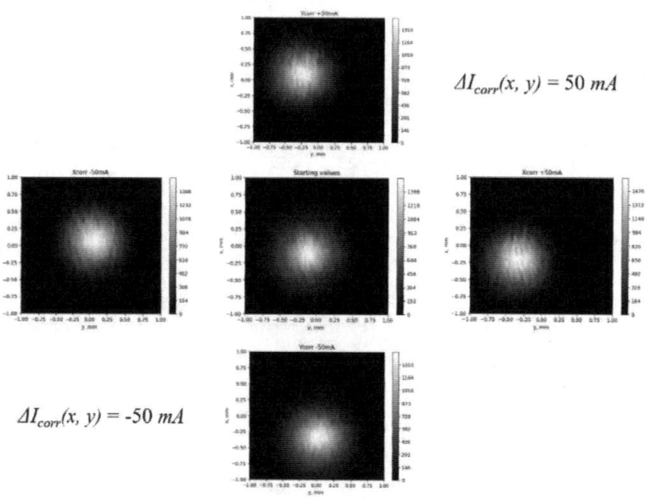

Figure 4. Magnetic corrector values selection example

4.4. Results and discussion

The results of measuring the beam current density obtained with the presented diagnostic system give more detailed information about the beam shape than the data obtained previously on the slit detector [2]. The modified data collection system and the visualizing software simplify the observation of the beam current distribution shape, shorten the setup time and increase its accuracy. As a result, obtaining a better beam quality became possible at the installation.

Currently, the theoretically calculated beam shape (see Figure 5) cannot be distinguished with sufficient accuracy from the image of the beam shape obtained with this diagnostic system. However, the observed similarity of the distribution character indicates that the theoretical predictions were correct. Moreover, the previously observed shape of the defocused beam on the luminophore is visually similar to the projections of an under focused beam obtained by the diagnostic system.

IOP Conf. Series: Materials Science and Engineering **759** (2020) 012005 doi:10.1088/1757-899X/759/1/012005

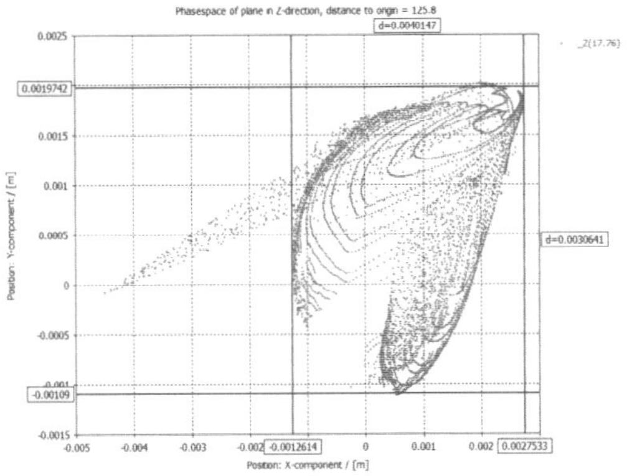

(a)

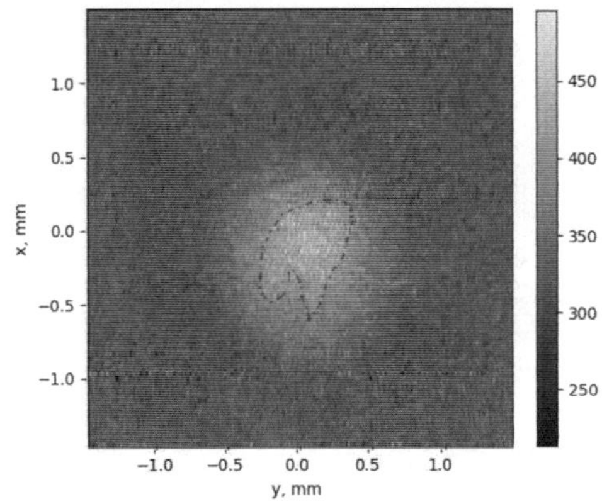

(b)

Figure 5. Beam current distribution (axes are co-directed): (a) – simulated at the focusing point by CST; (b) – measured at the welding point.

The principal defect of the system is the operation speed. The complete measurement cycle without iterations takes 2 minutes and that one with reducing noise iterations takes 7-10 minutes. Subsequently the scanning time must be reduced by 1-2 orders.

Image quality is strongly influenced by noise. Not all the noise can be eliminated by increasing the number of iterations. The installation sources of noise are:

- equipment that controls the beam current stabilization by high-frequency grid voltage regulation;
- deflecting system supplies stabilizing the coil current;
- high-voltage power supply stabilization system;
- heating current control equipment.

Only the beam current stabilization can be programmatically disabled at the control program user interface by putting the gun into non-stabilization mode with manual exposure of the grid voltage. All the data presented in this paper were obtained with this mode. Due to the noise that prevents the correct data collection, the results of the beam cross-section power density calculation were not presented in this paper. The calculated order of the power density is known to agree with that expected for the EBW beam.

5. Beam position and beam incline angle measurement system

5.1. Problem statement

We were tasked to develop a universal device for beam position and incline angle determination. This device should be using on electron beam welding facility, and other facilities where we need to know these parameters of electron beam (for example, on the facilities with alfa-magnet developed in our Institute [3]).

5.2. Construction of beam position and beam incline angle measurement system

To solve the problem a small diagnostic device located in electron beamguide was proposed. It displacement could be both closed to anode to check initial spreading of EBW electrons and downstream from the magnetic system, for example, for alfa-magnet facility. The diagnostic system consists of two independent deflecting coils (two-coordinate corrector) and electron collectors located further in the beam transition system. The center region of the device has an opening for a beam passage. The scheme of the device is illustrated on Figure 6.

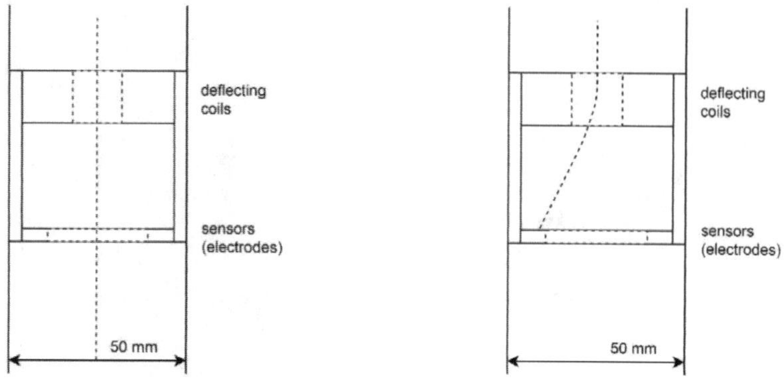

Figure 6. Scheme of beam position and beam incline angle measurement system.

The electrons collector consists of several lamellas located nearby electrically connected to ADC. Each lamella signal is measured separately of another (see Figure 7a).). The lamellas are proposed to be arranged in groups of 4 mirror-symmetrical on both sides, for each of the deflecting system axes. There are 16 electrodes for entire system.

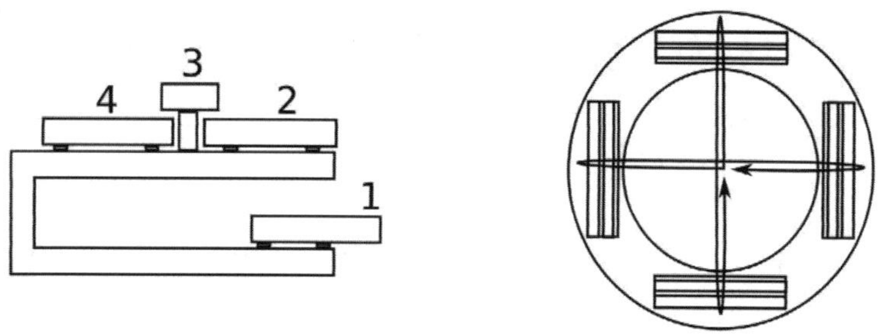

(a) (b)

Figure 7. Scheme of the device: (a) – lamellas scheme; (б) – scanning scheme.

Deflecting coil declines the electron beam along both diameters. When the beam is shifted or tilted, the different currents get to opposite electrodes, both in magnitude and duration. The scanning scheme is shown on Figure 7b. By the difference in the shape of the signals and by which lamellas the more current is detected, the beam inclination or displacement could be determined. To achieve the required accuracy of measurements, the transverse size of the lamellas should be on the order of the beam size (about 1 mm).

5.3. The analysis of the signals from the device

In order to avoid collector lamellas destruction due to heating by the beam the scanning procedure time should be as short as we can. At the moment we didn't calculate the optimal ratios between ADCs, DACs, power amplifiers for the magnetic system performance, the size and raw material of electrodes, and maximum current for the scanning procedure.

The Figure 8 demonstrates the qualitative dependence of expected signals from the lamellas. Axis x describes the coordinate in the space that corresponds to the magnetic system current. Axis y describes a beam current, which got on the lamellas.

On the Figure 8a is shown the case if the beam propagates the beamguide along the axis without a bias or tilt. The signals are symmetric and equal.

When the beam is biased on Δx from the beamguide center, the result of beam scanning procedure will look like the Figure 8b. Middle point between two corresponding signal peaks indicates the beam shift from deflecting lens magnetic axis.

IOP Conf. Series: Materials Science and Engineering **759** (2020) 012005 doi:10.1088/1757-899X/759/1/012005

The beam inclination results in non-equal shapes of the signals, as shown at the Figure 8c. It seems to be possible to calculate the beam incline angle analyzing the signal shapes. The exact formula of this calculation is determined by the geometry and location of electrodes. Moreover, the device is able to show if the beam is defocused (for example, high overlay of signals from lamellas 2 and 4, or smaller signal level from the lamella 3 than signal level from lamellas 2 and 4, or more smooth edges of the signals.)

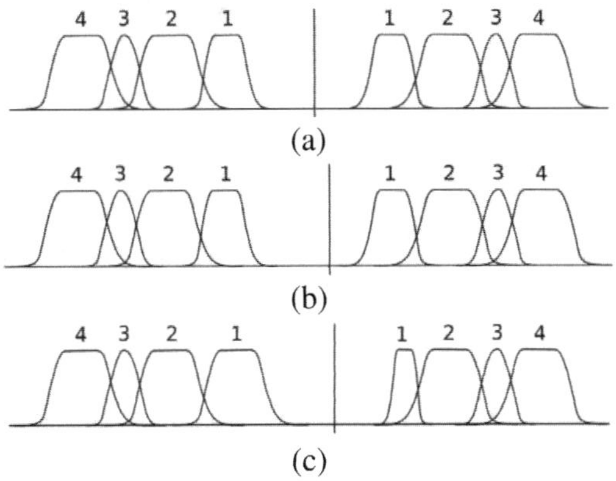

Figure 8. Signals from lamellas: (a) – straight beam with optimal focusing at the central position; (b) – beam displaced from the center; (c) – inclined beam.

5.4. Tests of the prototype

The prototype of the device for one magnetic corrector axis with copper lamellas (corresponding to electrodes 2 and 3 shown on Figure 7a) was manufactured and tested. During the preliminary test, there was set the function of slow beam oscillation with a beam current of 1.4 mA in the direction corresponding to the corrector axis. Signals from the lamellas were recorded by the oscilloscope (See Figure 9). The test was also carried out with no current stabilization mode. For extra noise reduction the RC-circuit was made for each electrode. The results obtained indicate the operability of such a device design. However, in this case additional measures are required to suppress high-frequency noise. In the next version of the prototype, the remaining electrodes need to be added in order to verify all of the described abilities, and the noise reduction scheme needs an improvement. As a result of the next series of preliminary tests, it is planned to collect the first database required for developing a user program to analyze signals and display the results for the beam parameters correction.

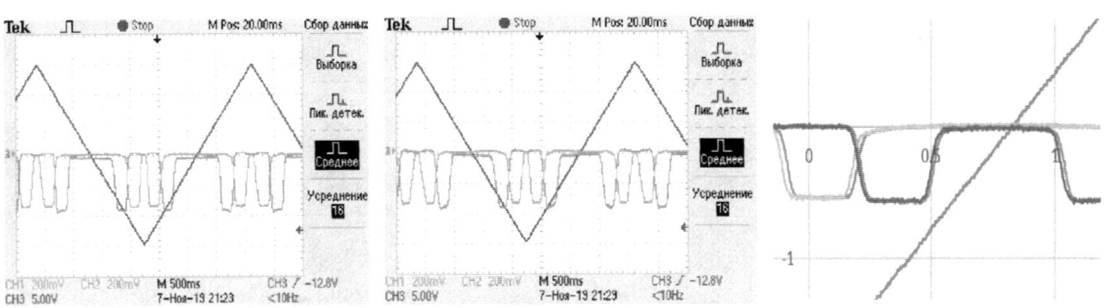

(a) (b) (c)

Figure 9. Signals obtained with the device prototype: (a) – straight beam with optimal focusing; (b) – unfocused beam; (c) – small angle inclined beam signals comparison.

6. Conclusion

The system of the beam cross-section current distribution measurement was put into operation at the EBW installation. The prototype of the system of beam position and beam incline angle measurement is at the testing stage. Measures were taken to decrease the scanning procedure time and there were identified further actions for minimizing this time. At the installation the works aimed at modernization of the equipment for accelerating ADC reading speed and decreasing the equipment noises are being carried out.

References

[1] Semenov Yu I, Logatchev P V et al. 2018 60 keV 30 kW electron beam facility for electron beam technology *Proceedings of EPAC08*. Genoa, Italy. TUPP161.
[2] Alyackrinskiy O N et al 2018 Prototype of source of electron beam with laser heating of the cathode *Nauchnoe Priborostroenie* **28**(4) 8-14.
[3] Alyackrinskiy O N et al 2019 Prototype of electron source with magnetic beam rotation for electron beam technologies *Nauchnoe Priborostroenie* **29**(1) 26-32.
[4] Sizov M M, Starostenko A A, Blokhina K A, Medvedev A M 2018 FPGA based image processing system for electron beam welding facility. *Proceedings of PCaPAC2018, Hsinchu, Taiwan.*

IOP Conf. Series: Materials Science and Engineering **759** (2020) 012006 doi:10.1088/1757-899X/759/1/012006

Electron-beam welding of large thickness parts in a narrow gap

I S Chulkov, A L Goncharov and A V Chulkova

National Research University "Moscow Power Engineering Institute", 14, Krasnokazarmennaya str., Moscow, 111250, Russia

E-mail: ChulkovIS@mpei.ru

Abstract. The paper presents a method of large thickness parts welding in a narrow gap with filler wire. A modernized deflection magnet is proposed, which allows deflecting an electron beam directly in the zone of weld pool. The distribution of magnetic induction in a local magnetic field along the axis of electron beam guide was measured and the results obtained were compared to the calculated ones. A photographic recording of electron beam deviation in a local magnetic field created by the presented deflection magnet was carried out. The angle of electron beam deviation about electron beam guide axis is estimated.

1. Introduction

When creating modern products, the use of different welding technologies is now widespread. For example, electron-beam welding (EBW) is one of the most essential in obtaining of one-piece joints made of stainless steels, titanium alloys, non-ferrous and refractory metals or their alloys. Nowadays electron-beam welding is widely used in different power engineering industries, aircraft manufacturing, shipbuilding and in cryogenic equipment production. Such a various application of this process is associated with a number of its advantages. The production of welded structures using electron-beam welding technologies allows obtaining welded joints with mechanical properties almost at the level of parent metal properties. However, welding of large thicknesses (over 100 mm) parts always presents a number of difficulties, one of which is increased probability of defects formation and complexity of their correction. In addition, in this case, high demands are made on the power of the equipment used [1, 2].

The problem of obtaining high-quality welded joints during butt welding of large thickness materials is associated with the possible appearance of various defects, such as incomplete fusion due to electron beam deviation from the joint axis, defects formation in weld root, sagging of weld root, formation of undercuts, pores and cavities. The reason for these defects formation can be such factors as inaccurate beam guidance on the junction, instability of technological parameters during welding process, excessive heat input and some others [1-3].

The paper presents a method of large thickness parts electron-beam welding in a narrow gap using filler material (wire) to improve weld formation and reduce the probability of most these defects formation. Such approach allows combining the positive qualities of layer-by-layer welding and special features of electron-beam welding. Using this method, it is possible to carry out layer-by-layer control of welding process, create an optimal source of heating, and vacuum protection allows realizing high-quality welding of chemically active metals.

IOP Conf. Series: Materials Science and Engineering **759** (2020) 012006 doi:10.1088/1757-899X/759/1/012006

2. Description of proposed method

The proposed electron-beam welding method was tested on the ELA-15I plant. To realize this method, the deflection magnet used for electron beam deflection was modernized. For its making it was proposed to use a material with a higher magnetic permeability (09G2S steel (C~0,09%, Mn~2% Si~1%, Fe balanced) in the annealed state) compared to that used previously (EP 517 steel (C~0,16%, Mn~0.5%, Cr~12%, Si~0.5%, Ni~2%, Mo~1.6%, W~1%, Nb~0.3%, Fe balanced), because it was necessary to obtain larger values of magnetic induction (more than 10 mT) for realization the method of welding in a narrow gap. Such values were not possible to obtain with a deflection magnet made of EP 517 steel and used earlier.

Typical deflection magnets do not allow electron beam reaching directly the material treatment zone during the process of welding materials of large thicknesses in a narrow gap. This is due to the fact that electron beam does not enter the weld pool zone, but is scattered on the side walls of the parts being welded, and when they are heated, a significant part of its power is lost. This leads to a change in the beam parameters and disruption of the entire process as a whole.

Therefore, to ensure the stability of joints formation during electron-beam welding of large thicknesses parts in a narrow gap and to eliminate defects formation in the weld and fusion zone, the deflection magnet was developed, which allows deflecting electron beam in close proximity to weld pool for obtaining a heat source of complex shape (figure 1).

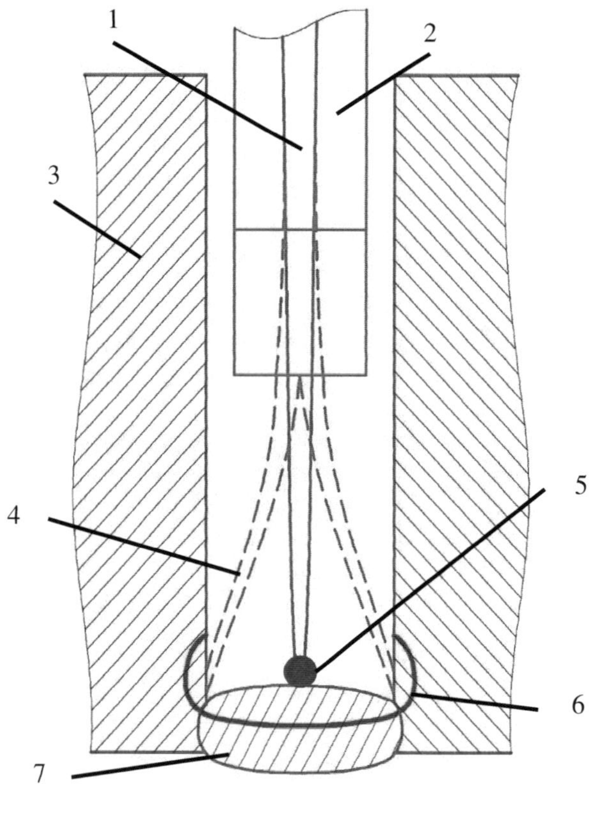

Figure 1. Scheme of electron-beam welding in a narrow gap: 1 – electron beam; 2 – deflection magnet; 3 – parts to be welded; 4 – electron beam extreme positions; 5 – filler material; 6 – scheme of melting isotherm; 7 – weld.

Design and operating parameters of the deflection magnet were selected using mathematical modeling of the process of electron beam deflection in a local magnetic field [1]. The Finite Integration Technique (FIT) [2] developed by Weiland in 1977 was used for the calculation. This numerical method provides a universal spatial discretization scheme applicable to various electromagnetic problems according to the equations

$$\oint_{\delta A} \vec{H} \cdot d\vec{s} = \int_{A} \left(\frac{\delta \vec{D}}{\delta t} + \vec{J} \right) \times d\vec{A}, \quad (1)$$

$$\oint_{\delta A} \vec{B} \cdot d\vec{A} = 0, \quad (2)$$

where $\vec{H}$ is magnetic field strength, A/m; s is closed contour ($d\vec{s}$ is contour element, m); $\vec{D}$ is electric induction, C·m^{-2}; t is time, s; A is the surface stretched over the contour s ($d\vec{A}$ is vector of area element, directed outward from the considered volume); $\vec{B}$ is magnetic field induction, T.

Tracking electron's movement is based on the discretization of electric and magnetic forces by means of the following equations:

$$m\frac{d\vec{\upsilon}}{dt} = q\left(\vec{E} + \vec{\upsilon} \times \vec{B}\right), \quad (3)$$

$$m^{n+1}\vec{\upsilon}^{n+1} = m^{n}\vec{\upsilon}^{n} + \Delta t\left(\vec{E}^{n+1/2} + \vec{\upsilon}^{n+1/2} \times \vec{B}^{n+1/2}\right), \quad (4)$$

where m is particle mass, kg; q is its charge, C; $\vec{v}$ is speed vector, m/s; $\vec{E}$ is electric field vector, V/m; dt is discrete time step, s; n is index of time step, at which the parameter is determined (speed, electric intensity and induction).

In the calculation, vacuum was defined as the medium for electron motion. In order to determine the magnetic induction in the gap of deflection magnet made of 09G2S steel, a magnetization curve was set, from which the magnetic permeability was determined depending on the magnetic field strength. The electron beam in the calculation has a fixed current value, which is equal to 100 mA. The main input data for the deflection magnet parameters selection were current and number of turns in inductors, as well as magnetic properties of the materials, specified in the form of magnetization curves.

Scheme of the proposed method of large thickness parts welding in a narrow gap with filler wire is shown in Figure 1. A deflection magnet is placed in a narrow gap between the parts to be welded of non-magnetic material 3. An alternating magnetic field, which deflects electron beam to position 4, is created in the magnetic gap between the tips. To fill the technological gap between the parts to be welded 3, a filler wire 5 is fed, which is melted by electron beam, when its axis coincides with the axis of deflection magnet. Electron beam deviation in the zone of weld pool is carried out to ensure the stability of joints formation during electron-beam welding of large thicknesses parts in a narrow gap and to eliminate defects formation in the weld and fusion zone. Thus, to obtain a more even temperature isotherm 6, a heat source is created that acts on the walls of weld joint and surface of the weld.

Other authors previously obtained results related to electron beam deflection in an electrostatic field in in close proximity to treatment zone. This made it possible to expand the technological capabilities of EBW [4-7], but using of an electrostatic mirror requires to select the shape of reflecting surface individually for different shapes of joints. In addition, the application of this method is difficult during welding in a narrow gap and does not allow use of large beam currents.

To determine the magnitude of resulting magnetic field and the nature of its distribution, magnetic induction was determined along electron beam guide axis [8-12]. Its measurement was carried out according to the scheme shown in figure 2, using the Hall sensor (model AD2251). During the measurement of magnetic induction, the current in the inductance coils was recorded (its values are presented in the explanation to figure 3). The obtained current values were used for numerical simulation of electrons motion in a magnetic field.

To fix the deflection trajectory of electron beam by deflection magnet, there was used method of recording the glow of residual gases in the visible range of spectrum [13].

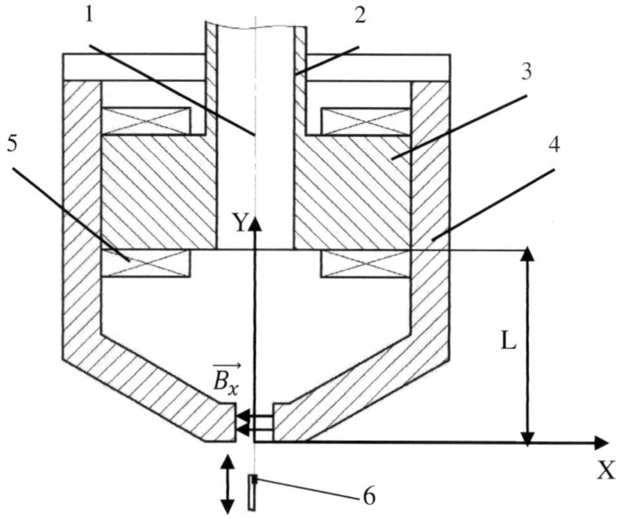

Figure 2. Scheme of magnetic induction measurement: 1 – beam axis; 2 – electron-beam guide; 3 –magnetic core; 4 – magnetic tip; 5 – inductor; 6 – Hall sensor.

3. Results of studies

The distribution of magnetic induction in the gap between magnetic tips was obtained at various current values. The distribution of magnetic induction along electron beam guide axis in deflection magnet is shown in figure 3. Figure 3 shows the experimental data and calculated values obtained using numerical simulation. It can be seen from the graphs that the curves have a similar form and, with increasing current strength, the experimental data have values higher than the calculated ones.

During the tests, photographs of electron beam deviation in a local magnetic field were obtained (figure 4).

Using the characteristic trace of electron beam on the target, the deviation angle from the axis of electron-beam gun was further determined; it was approximately 13°. According to the calculated data, the deviation angle was 11°.

4. Conclusion

The results of deflection magnet testing have good repeatability. The graphs of magnetic induction distribution in the gap have a similar form. The data obtained experimentally are slightly larger than the calculated ones. This fact can be explained by the presence of unaccounted parameters in the calculation, such as external magnetic field and structural features of the material.

The experimentally obtained value of inclination angle for electron beam corresponds to the calculated value. Since the magnetic deflection system is designed to operate in a narrow gap of about 6 mm, the obtained declination values are sufficient for realization of the technological process according to the scheme shown in figure 1.

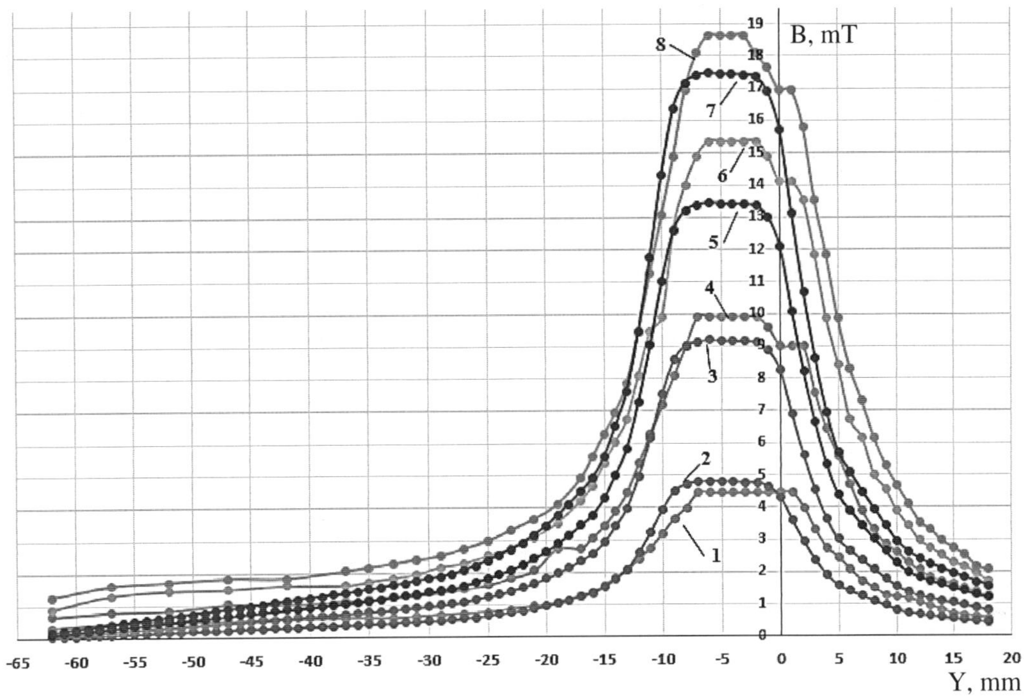

Figure 3. Distribution of magnetic induction along electron beam guide axis in deflection magnet. Calculated values: 2 – current is 0.123 A; 3 – current is 0.234A; 5 – current is 0.342A; 7– current is 0.444 A. Experimental data: 1 – current is 0.123 A; 4 – current is 0.234 A; 6 – current is 0.342A; 8 – current is 0.44 A.

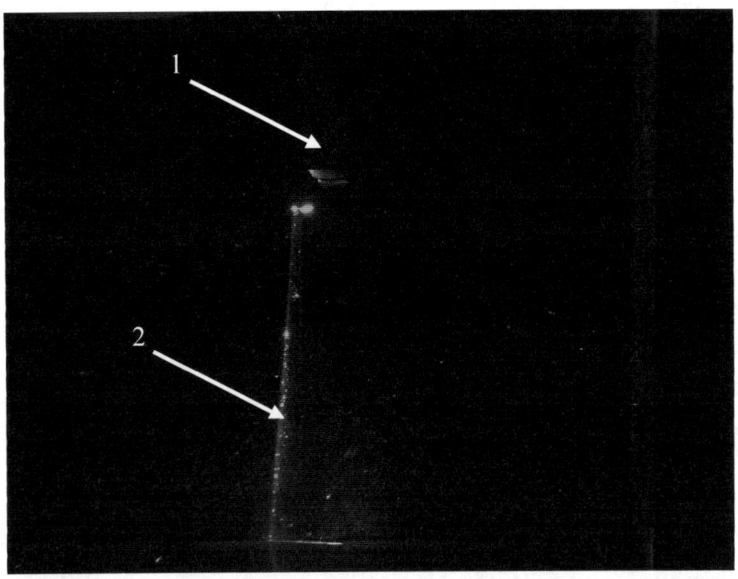

Figure 4. Photography of electron beam deviation in a local magnetic field: 1 – magnetic tips; 2 – electron beam, deviated in a magnetic field.

5. Acknowledgments
The research was carried out in National Research University "Moscow Power Engineering Institute" with the financial support of the Russian Science Foundation (project No. 18-19-00652).

References

[1] Chulkov I S, Goncharov A L and Sliva A P 2018 *Journal of Physics: Conf. Series 9. "9th Int. Conf. on Beam Technologies and Laser Application, BTLA 2018"*. **1109(1)** 012005
[2] Weiland T. A 1977 *Electronics and Communication, (AEÜ)* **31** 116-20
[3] Nazarenko O K, Kaydalov A A and Kovbasenko S N 1987 *Electron-beam Welding* (Kiev: Naukova Dumka) p 256
[4] Zuev I V 1998 *Processing Materials with Concentrated Energy Flows* (Moscow: Publishing house MPEI) p 162
[5] Drovosekov A V and Verveiko D S 2018 *Actual Problems of Aviation and Astronautics* **1(14)** 429-31
[6] Borovik V M, Frolov O V and Shubin F V 1991 *XI All-Union Scientific and Technical Conference on Electron Beam Welding (Nikolaev). – Shipbuilding* 13-5
[7] Grigoryev G V, Kasitsyn A N and Sliva A P 2015 *Int Conf on Electron Beam Welding and Related Technologies, MPEI (Moscow)* 374-87
[8] Scherbakov A V, Goncharov A L and Chadova I I 2012 *Vestnik of (National Research University MPEI (Moscow)* **2** 85-90
[9] Goncharov A L, Dragunov V K, Shcherbakov A V, Portnov M A and Chulkov I S 2015 *Russian Electrical Engineering* **86(10)** 594-97
[10] Scherbakov A V, Rodyakina R V, Kozhechenko A S, Gaponova D A, Goncharov A L and Dragunov V K 2017 *Let. to the J. of Tech. Phys.* **43(21)** 10–5
[11] Novokreshchenov V V, Rodyakina R V and Karimbekov M A 2017 Solid State Phenomena. **265** 237-45.
[12] Shcherbakov A V, Martynov V N, Kharitonov I A, Gaponova D A, Rodyakina R V and Dragunov V K 2018 *Russian Electrical Engineering* **89(4)** 249–54
[13] Goncharov A L, Sliva A P, Terentyev E V, Portnov M A and Chulkov I S 2014 *Electrotechnica & Electronika "E + E". 11th Int Conf on Electron Beam Technologies (Varna, Bulgaria)* **108** 259-63

Creation of corrosion resistant coatings on ductile iron by re-melting flame sprayed layers using laser and electron beam

D Drimal [1], **F Kolenic**[1], **M Simek** [1]

[1] PRVA ZVARACSKA, a. s., Kopcianska 14, Bratislava, 851 01 Slovak Republic

E-mail: drimal.daniel@pzvar.sk

Abstract. The paper presents the results of the research of the formation of thin surface corrosion resistant layers on ductile iron with spheroidal graphite. Surface layers were created by a two-step technological process. In a first step, the surface layer material was applied to the surface of the ductile iron by means of a flame-powder coating in a thickness of 0.8 to 1 mm. The treatment of the layer was carried out in the second step by means of a defocused laser beam or a programmed-deflected electron beam. The influence of the laser beam defocusing and the influence of the energy distribution of the deflected electron beam on the characteristics of the formed thin surface layer was studied. 40 mm thick GGG 40 ductile iron samples were used as the substrate experimental material. The Ni-B-Si-type nickel powder with the designation NP22 and Ni-Cr-B-Si-type nickel powders of the type NP52 were used to form the surface layer. The quality of layers was assessed on the basis of surface forming, metallographic evaluation of selected properties of the layer and measurement of micro-hardness at the interface of the layer and the substrate material.

1. Introduction

In modern history, the first ductile cast iron was produced in 1934, but its broader industrial use is dated from the discovery of possibility of its modification by magnesium in master alloy with nickel in 1942.[1]. Extremely good mechanical properties, good workability and relatively low price per kilogramme are the most important reasons for a broad use of ductile cast irons in industry. An alloy with graphite in the form of globules is characterised by a strength of 400 MPa and has approximately 10% lower density in comparison with cast steel. It is used mainly in the automotive and mechanical engineering industry. As the nuclear energy industry was developing, it also became widely used in the production of heavy thick-walled castings – transport and storage containers for spent nuclear fuel [2-5]. Due to a relatively low corrosion resistance of ductile cast iron, the need for a significant increase of its corrosion resistance arises. One of the solutions to this problem is creating a thin surface layer on functional and contact surfaces of the container. General requirements for surface finishing of functional surfaces of a cast iron container intended for use in nuclear energy industry are as follows:

- High corrosion resistance.
- Provision of surface decontamination (dependent on surface porosity).
- Heat resistance of the layer ca. up to 300°C.
- Stability in radioactive environment.
- Resistance to potential damage during handling.
- Homogeneity of the layer also after metalworking without through discontinuities.
- Guaranteed cohesion with the base material.
- Repairability.

- Inertness to a boric acid solution in the case of wet storage of spent nuclear fuel.

Similarly to surface finishing of other metallic materials, arc and beam technologies are applied to cast iron in creating thin, corrosion resistant surface layers. However, the problematic weldability of cast iron prevents practical applicability of these technologies. The greatest obstacle is a high content of carbon, which causes the thermally affected area to be prone to opacification. Also, the situation is complicated due to the ability of solidification of cast iron according to Fe-Fe_3C metastable system, when C is in the structure in the form of ledeburite and cementite, which can have a significant impact on the solidity of the created layer [1, 2]. The objective of the presented experimental tests was to design and verify the technology of creating thin corrosion resistant layers on the ductile cast iron with globular graphite by applying remelting of sprayed flame-powder layers using laser and electron-beam technologies.

2. Problem analysis

Nowadays patented methods of electrolytic metallic coating are used for creating functional surface layers on cast irons. The created layers are corrosion resistant and relatively easily decontaminable. However, they possess low resistance to mechanical damage. The methods of creating layers of required properties on cast irons with globular graphite using welding technologies have significant difficulties. The requirements connected with reaching high level of adherence of the layer to the base material and with resistance of the layer to deterioration demand creation of a good metallurgical bond between the layer and the base material. On the other hand, the requirement for a good corrosion resistance poses challenges for a low level of blending the material of the layer with the base material.

Due to the high carbon content, the base material (ductile graphitic cast iron) requires a specific heat mode in the process of surfacing. There is an opinion accepted by professionals that during surfacing of cast iron using arc technologies, application of pre-heating and post-heating is required, with the aim to lower the steepness of thermal gradients and provide solidification of the cast iron according to the stable system (Fe – graphite). The possibility to use arc methods for creating corrosion resistant and decontaminable layers on the ductile cast iron was researched in numerous works [2,5]. Several noteworthy results were achieved within the research. For example, the use of TIG pulse mode with pulse frequency in the order of hundreds Hz enabled to create solid single-layer- or multiple-layer claddings with nickel-based additive material without the need to apply pre-heating and/or post-heating. The disadvantages are proneness to cracking as well as low productivity [4].

It appears that application of beam technologies, such as laser and electron beam, is a suitable technological solution. The main assumption for the use of beam technologies is the fact that they have sufficient power to create a metallurgical bond at the interface of the layer with the base material. They enable to create 2D heat fields by defocusing of a laser beam or electron beam and create conditions for minimum blending of the base material with the material of the layer. In addition, during application of electron beam, we can perform the process in vacuum that provides perfect protection of the process against oxidation and also enables degassing of the fused layer. By using a programmable biaxial deflection of electron beam, the conditions for optimisation of areal distribution of beam energy in the process of sprayed layer processing are created [6].

3. Experimental material

The claddings were executed on test samples with the dimensions of 100 x 100 x 40. The samples were made of GGG 40 alloy type with globular graphite. Following technologies were used:

1) Remelting of the sprayed flame-powder layer by defocused laser beam.
2) Remelting of the sprayed flame-powder layer by programmed electron beam deflection.

The quality of layers or claddings was assessed based on:

- Visual inspection of surface formation of the cladding.
- Metallographic inspection of integrity and structure on cross sections of claddings (base metal – transition layer – cladding). The samples of cross sections of claddings were mechanically ground and polished, and etched for improving the contrast of the structure. The observation

was performed using an OLYMPUS GX 51 light microscope equipped by an ARTCAM 300MI digital CCD camera.

- Micro-hardness measurements – at the interface of the cladding testify to the character of bond and size of blending of the filler material with the base. The test was performed according to STN EN 1043-2 with a load of 100 p and a load time of 10 s. Micro-hardness was measured by a Haneman micro-hardness meter connected to a NEOPHOT2 metallographic microscope.
- The IPG YLR 4500 solid-state fibre laser with the maximum output power up to 4.5 kW was used as a laser source. The laser beam was transported to the PRECITEC YW50 welding technological head using optical fibre with a diameter of 0.3 mm. The focal length of the focusing lens was 200 mm. Protection of surface in laser remelting was provided by an inert gas.
- As a source of electron beam, the PZ EZ ZH4 electron-beam welding device was used, with the power of electron beam up to 15 kW and accelerating voltage of 60 kV, equipped by a system of dynamic biaxial deflection of PZ DF 5 type.

In **laser remelting** of the spraying, a defocused laser beam with a wave length of 1.06 μm was used. The IPG YLR 4500 solid-state fibre laser with the maximum output power up to 4.5 kW was used as a laser radiation source. The laser beam was transported to the PRECITEC YW50 welding technological head using optical fibre with a diameter of 0.3 mm. The focal length of the focusing lens was 200 mm. During the processing, the focus of laser beam was set in the distance from +40 to +60 mm above the sample surface, which ensured the beam defocusing. By laser beam defocusing, the beam energy concentration is changing and the energy interacts with the sprayed powder layer. Protection of surface in laser remelting was provided by an inert gas. NP 22 powder was used in the laser experiments.

In **electron-beam processing**, the remelting of the spraying was performed on the PZ EZ EZ4 electron-beam welding device was used, with the power of electron beam up to 30 kW and accelerating voltage of 60 kV. The PZ DF 5 system intended for programmed dynamic biaxial deflection was used for remelting. The biaxial, computer-controlled deflection system enabled to change the area distribution of electron beam by programme. An advantage of electron beam is the fact that the remelting process is carried out in the high vacuum environment, which prevents oxidation, while ensuring good degassing with the effect of decreasing porosity of the remelted layer. A disadvantage is the need of creating a vacuum environment, which may increase the expenses per area unit of cladding creation.

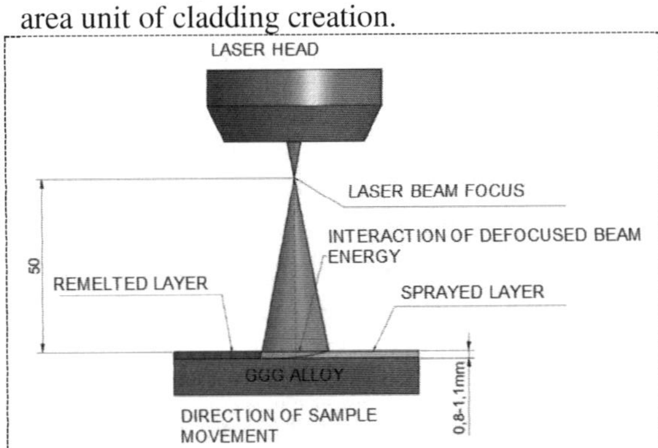

Figure 1. Principle of laser remelting of the sprayed layer

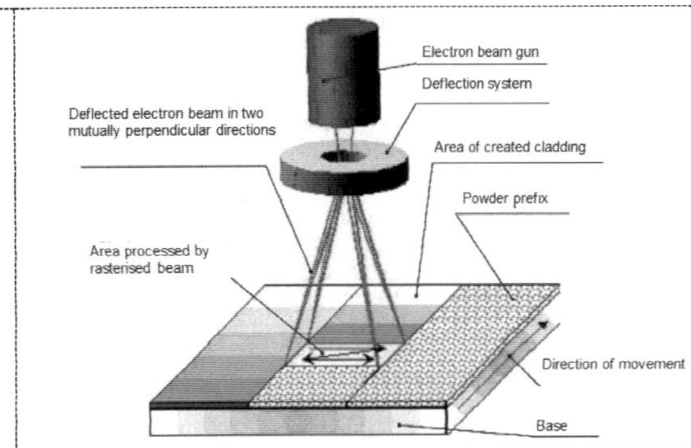

Figure 2. Principle of processing (remelting) of the sprayed layer by electron beam.

The principles of surface processing of the layer applied to the surface of the base material by the technology of flame-powder spraying are depicted on figures 1 and 2. The areal distribution of laser beam energy has a static Gaussian distribution. On the other hand, the PZ DF5 programmable system of electron beam deflection enables the beam deflection in the angle of ±7° in the latitudinal and also longitudinal direction to the surfacing movement. The maximum frequency in both deflection axes is

15 000 Hz. By programming the deflection system, we influence the electron beam energy distribution upon the surface layer, i.e. in the processing of the layer. The system enables to reach a constant energy density during the layer processing or any course of beam energy density based on programme commands.

4. Experimental material

4.1. Base material

Cast iron GGG 40

Ductile cast iron (EN-GJS-400-18) is ferritic with C in the form graphite globules with a minimum tensile strength of 400 MPa. It features high shock resistance and relatively good thermal conductivity. The reference chemical composition of the alloy is specified in Table 1.

Table 1. Chemical composition of cast iron GGG 40 in wt. %

C	Si	Mn	S	P	Mg
3.4 – 3.6	2.5 – 2.7	Max. 0.2	Max. 0.01	Max. 0.05	0.03 – 0.08

4.2. Material of applied layer

NP 22 powder

Ni-B-Si-based nickel powder marked as NP22 was used for flame-powder spraying. The NP 22 powder is slag-forming, intended for thermal spraying. The particles have a spherical shape and size of 45 – 90 μm. The melting temperature of NP 22 powder is 1 100 °C and the applied layer reaches hardness of 18 – 24 HRC. The applied layer is corrosive resistant, abrasion resistant and well machinable. The reference chemical composition of NP 22 alloy is specified in Table 2.

Table 2. Chemical composition of powder NP 22 in wt. %

C	Si	B	Fe	Ni
Max. 0.25	Max. 5.0	Max. 3.0	Max. 5.0	Rest.

NP 52 powder

Ni-B-Si-Cr-based nickel powder was used for flame-powder spraying. The NP 52 powder is slag-forming, intended for thermal spraying. The particles have a spherical shape and size of 45 – 90 μm. The melting temperature of NP 52 powder is 1 000 – 1 100 °C and the applied layer reaches hardness of 50 – 54 HRC. The applied layer is flame and corrosive resistant and well machinable. The reference chemical composition of NP 52 alloy is specified in Table 3.

Table 3. Chemical composition of powder NP 52 in wt. %

C	Si	B	Fe	Cr	Ni
Max. 0.4	Max. 5.0	Max. 3.0	Max. 5.0	Max. 11.0	Rest.

5. Experimental results

5.1. Lase remelting of flame-powder spraying

Within the experimental works carried out, the possibility of two-stage creation of surface layer with special properties on the ductile cast iron. A layer of Ni-based powder (NP 22, NP 52) was applied to the base material using the flame-powder technology. The thickness of the spraying was 0.8 – 1.2 mm. The structure of the sprayed layer is characterised by a relatively high porosity and a highly rough surface, which hinders the possibility of decontamination or makes it impossible, and lowers the corrosion resistance. Figure 3 depicts the surface of the sprayed layer and Table 4 presents the results of areal semi-quantitative analysis of presence of the most important alloying elements of the layer, sprayed by flame-powder spraying. The layer was sprayed by NP 52 nickel powder. The areal EDX analysis results show that the chemical composition corresponds to the reference chemical composition of the powder used. The measured value of the C content is distorted with errors caused by possible contamination with organic compounds during manipulation and preparation of the sample. The issues of the EDX method in measurement of the content of light atoms is next source of the errors.

Table 4. Results of areal EDX analysis of the spray layer

	C	Al	Si	Ca	Cr	Fe	Ni	Total
Spectrum	29.58	1.02	3.38	0.59	10.13	7.16	48.14	100

The measured results point out the fact that the properties of the sprayed layer do not comply with the requirements for corrosion resistance and decontaminability. For this reason, the surface layer was further processed by remelting by laser and electron beam.

In the process of remelting by laser, a defocused laser beam that enables creation of a wider laser beam spot. The defocusing parameters were defined in a way that the laser beam focus was in the distance of +40mm, +50 mm and +60mm above the layer surface. NP 22 nickel powder was used as layer material. With the laser beam power of 4.5 kW, we defined the defocusing value of +50 mm above the surface of sprayed layer as optimum. The width of the laser beam spot was 9 mm. The parameters of processing (remelting) are specified in Table 5. Surface of the remelted layer was smooth without visible surface pores or non-integrities (figure 3).

Table 5. Results of areal EDX analysis of the spray layer

N.o	Laser beam power, kW	Velocity, mm/s	Focal position	Shielding gas flow rate, l/min
1.	4.5	7	+ 50	15, He
2.	4.5	5	+ 50	15, He
3.	4.5	5	+ 50	15, He
4.	4.5	5	+ 50	15, He

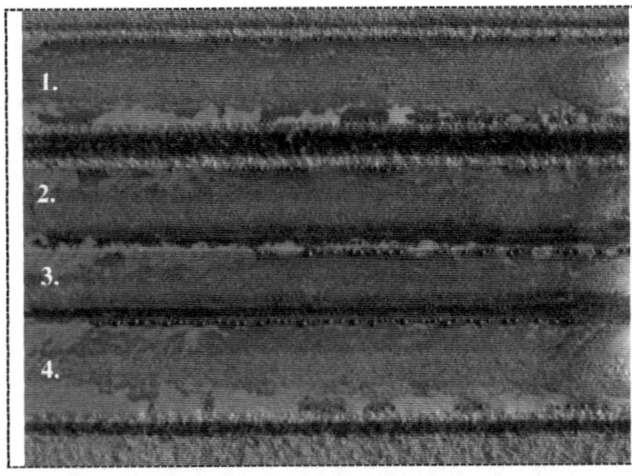

Figure 3. Surface of remelt runs made by defocused laser beam.

Since the powder contains a flux, a thin, inconsistent layer of slag was created on the surface of the cladding. Inconsistent distribution of energy in the beam was eliminated by applying a change of overlapping of individual weld beads.

A cross section of the remelted layer (NP 22) together with the details of individual cladding areas are depicted in Figure 4. Remelting of the powder spraying is not even throughout the width. The size of the fused area of the alloy is dependent on geometry of the defocused beam and thickness of the applied layer. The presence of closed pores was found in the cladding after remelting. It is possible to see a sintered character of an insufficiently remelted surface layer in the overlapping areas. By optimising the size of overlapping of individual trajectories of layer processing, it is possible to reach the required quasi-homogeneous densification of the sprayed layer with the good metallurgic bond with the base material.

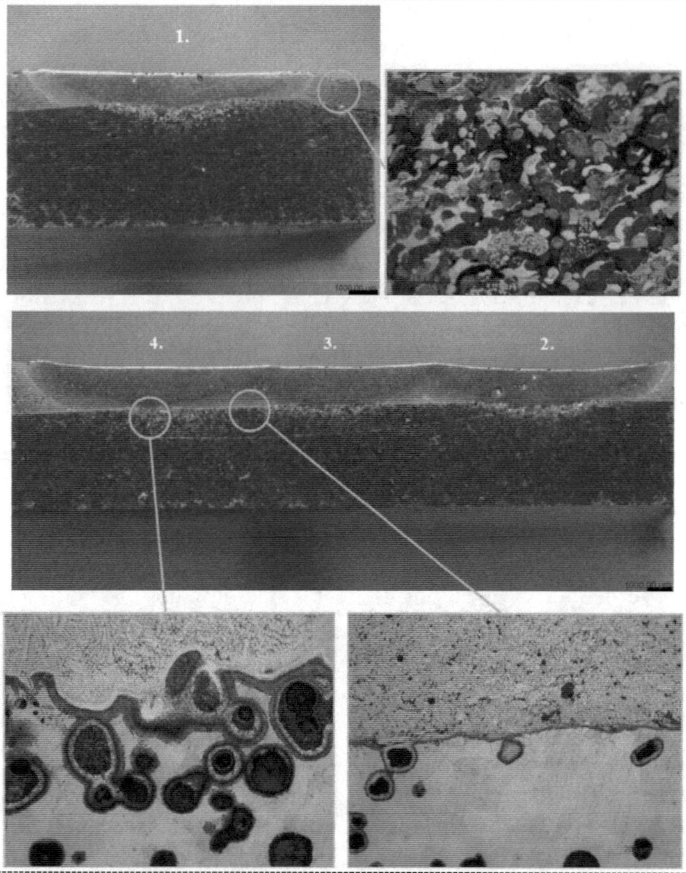

Figure 4. Structure on cross-sections of laser remelted layer

The micro-hardness measurement results are specified in Figure 5. The fusion line showed a moderate undulation. The depth of penetration of the melted material into the base material reached the value up to 200 μm. The hardness in the area of moderate blending ranged from 230 to 956 $HV_{0,1}$. The undulation of the interface is caused by the presence of graphitic nodules that melt down due to heat mode, which leads to creation of a eutectic with an increased hardness of 839 – 956 $HV_{0.1}$. In the heat-affected zone, during the heating beyond the austenitisation temperature, carbon saturation of the matrix around the graphite nodules occurred. In the subsequent cooling, the cooling velocity prevented a complete reverse diffusion of carbon into the graphitic particles and led to pearlite transformation.

In the transition areas where the beam energy was only enough for a only sintering of the powder, the undulation of the fusion border is lower and the creation of a eutectic was suppressed, which is proved by results of metallographic observation.

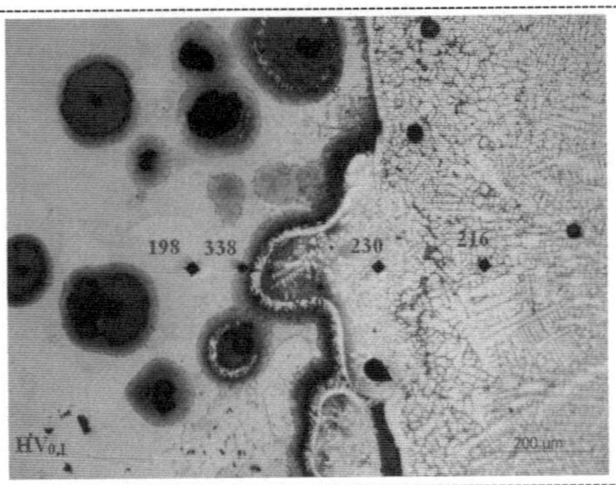

Figure 5. Micro-hardness at the interface cast iron GGG 40 – laser remelted layer (NP22 powder)-

5.2. Electron beam processing (remelting) of flame-powder spraying

A similar procedure of remelting of the sprayed layer was selected in the case of using electron beam. Electromagnetic biaxial computer-controlled deflection of electron beam was used for homogenisation of energy distribution in the beam spot. The remelting parameters are specified in Table 6.

Table 6. Electron beam remelting parameters

Process			Deflection	
Voltage, kV	Beam current, mA	Velocity, mm/s	Ch. A Amplitude / frequency	Ch. B Amplitude / frequency
55	60	5	linear 358 mV 30 Hz	Linear 3 583 mV 1.5 kHz

Figure 6. Surface of electron beam processed NP22 powder layer.

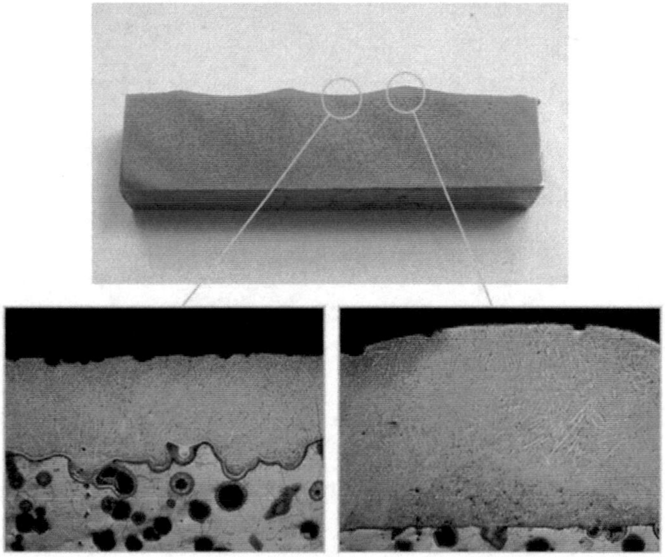

Figure 7. Cross section of a layer (NP22) melted by electron beam.

In the experiments, the samples of GGG 40 alloy sprayed by NP 22 and NP 52 nickel powders were used. Based on the visual assessment, the surface of the remelted layer was moderately undulating with excessive trace edges due to the melt pool dynamics during deflection (Figure 6). A moderate spatter on the surface of the claddings was observed as well.

A low blending of the welded layer with the base also resulted in lower values of micro-hardness in the fused area of the alloy (Figure 7). The structure of the remelted area was similar to the structure when using laser beam.

An areal energy – disperse analysis of the remelted powder spraying was created. depicts the surface formation of the sprayed layer of NP 52 powder and Table 7 presents the results of areal semi-quantitative analysis of presence of the most important metal parts of the layer. The EDX analysis results show that the chemical composition corresponds to the chemical composition of the powder used.

Table 7. Results of surface EDX analysis of electron beam melted (NP52) layer

	C	Si	Cr	Fe	Ni	Total
Spectrum	18.05	5.39	8.95	5.04	62.57	100

Figure 8 depicts the results of micro-hardness measurement. The hardness ranged from 272 to 667 $HV_{0,1}$. Due to a lower course velocity of remelting and areal deflection of the beam, a higher amount of heat was taken into the material, which led to lowering of the tendency to create hard structures in comparison with laser processing.

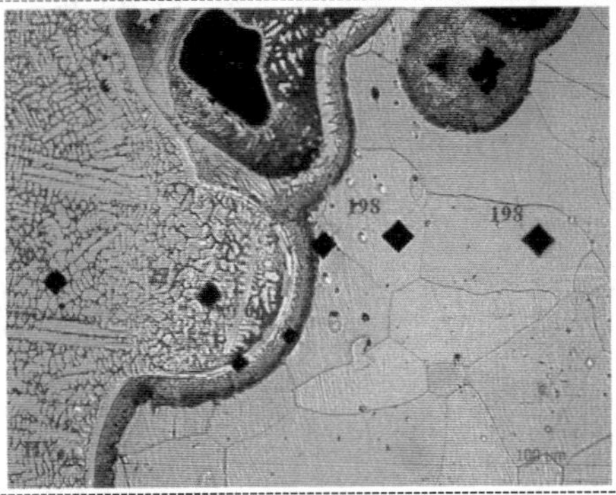

Figure 8. Micro-hardness measurement at the weld metal interface (NP22 powder) - cast iron GGG 40.

6. Conclusion

The objective of the submitted article is to present the possibilities of creating thin surface corrosion resistant layers on GGG 40 ductile cast iron with the use of beam technologies. Within the experimental research, the properties of the surface layers created by a two-stage technology using nickel-based powders were studied. A flame-powder spraying with the thickness of ca. 1 mm on the ductile cast iron was remelted by electron or laser beam. Experimental research brought following results:

- Surface layers remelted by laser as well as electron beam showed compact and without significant of surface.
- The physical principle of beam technologies enables precise regulation of energy density of the technological process. In comparison with other heat sources, it enables minimum impact on the welded base material.
- The surface of the remelted layer by electron beam was undulating due to the influence of the melting bath dynamics in rasterising, which may be suppressed by optimisation of pattern and parameters of electromagnetic deflection.
- A limitation of beam technologies is the size of thickness of the remelted layer per one pass. It was verified that an optimal thickness of the remelted layer in the fusion process of the ductile cast iron is within the range of 0.8 – 1 mm. In case of a greater thickness, the heat effect of the beam is insufficient for its complete remelting and a part of the layer remains sintered.
- The structure of the ductile cast iron influences the properties of the created layer. If graphite is not in the form of globules, but in the form of irregular shapes, a probability of more intensive dissolution of such particles increases, which raises the risk of formation of undesirable fragile and hard structures.

Based on the results of the presented research, we may state that application of beam technologies in the creation of corrosive resistant and decontaminable layers on ductile cast iron does not require pre-heating. To reach a solid cladding without the use of pre-heating, it is important to effectively dose the energy into the cladding process together with an appropriate selection of filler material composition (without carbide-forming elements, Cr, B) and thereby positively influence the heat-affected zone microstructure and/or minimise the amount of hard and fragile structures.

7. Acknowledgments

This research was supported by the Slovak Research and Development Agency based on Contract No. APVV-17-0432 as well as by the project of industrial research of the Ministry of Education, Science, Research and Sport within the call Stimuly 2018 based on Contract Reg. No. 1227/2018.

References

[1] Hrivňák I 2009 *Zváranie a zvariteľnost materiálov* (Welding and weldability of materials), Nakladateľstvo STU v Bratislave, , ISBN 978-80-277-3167-6.
[2] Pascual M, Ferrer C, Rayón E 2016 Weldability of spheroidal graphite ductile cast iron using Ni

/ Ni-Fe electrodes *International Journal of Scientific & Engineering Research* **7** 5.
[3] Hrivňák I 1992 *Theory of weldability of metals and alloys* Elsevier Science Ltd.
[4] Kearns W H & et al 1997 *Welding handbook*, Vol. 4, Metals and their weldability, 7 ed. Miami: AWS, ISBN 0-87171-218-1.
[5] Steiward W 2014 Experience with Ductile Cast Iron Fuel Casks, In *DSFM REG CONFERENCE*, Rockville, Maryland.
[6] Zenker R 1998 What's new in thermal surface treatment by electron beam?, In *Proc. of 6th International conference on welding and melting by EB and LB* (Toulon, pp. 527 – 534).

IOP Conf. Series: Materials Science and Engineering **759** (2020) 012008 doi:10.1088/1757-899X/759/1/012008

Research of thermoelectric effects and their influence on electron beam in the process of welding of dissimilar steels

A L Goncharov, A P Sliva, I A Kharitonov, A V Chulkova and E V Terentyev

National Research University "Moscow Power Engineering Institute", 14, Krasnokazarmennaya str., Moscow, 111250, Russia

e-mail:GoncharovAL@mpei.ru

Abstract. The results of studies of magnetic and thermoelectric properties of pearlite, martensitic and austenitic steels are given. The magnetization curves for the materials under study, as well as dependences of absolute thermoelectric power on the material temperature are obtained. The data of magnetic induction changing at the control points above the welded samples during EBW are achieved. The simulation of the magnetic field distribution during electron beam welding of dissimilar steels under the conditions of the thermopower currents generation is carried out. The shape of the electron beam being deflected in the magnetic field of thermoelectric currents is calculated. The calculated values are compared to the experimental results.

1. Introduction

The process of electron beam welding is characterized by a large number of physical phenomena occurring in the interaction of the accelerated electrons beam with the material, namely thermal, deformational, hydrodynamic and plasmic. Most of these processes are connected, moreover, the arising physical fields affect the processing tool – the electron beam – in response.

During dissimilar materials welding process the greatest influence on the electron beam is exerted by magnetic fields formed as a result of thermoelectric currents. Thus, the existing variety of physical processes can be supplemented by the thermoelectric phenomena, which under certain conditions becomes decisive in the formation of a high-quality welded joint. That is because the beam deflection caused by the magnetic field of thermoelectric currents can cause the lack of penetration in the thickness of the product and lead to a serious defect [1-7].

The value of the magnetic induction of thermoelectric current is determined by a number of factors including the temperature distribution in the welding zone, the shape and thickness of the welded part, as well as electrical resistance, magnetic permeability and thermo-EMF of the joined materials [3].

One of the main reasons for the appearance of thermoelectric power in dissimilar alloys is the deviation of electron system from the equilibrium state in the presence of a temperature gradient. Since in this case the driving force is a diffusion flow of a charge carrier in the field of temperature gradient this kind of thermoelectric power is called diffusion [8, 9]. The magnetic permeability of the material is not constant and depends on the magnetic field strength, as well as on the type of material and temperature.

Mathematical models of thermoelectric phenomena in electron beam welding have been developed to predict the value of the electron beam deflection from the joint [2, 3]. However, the assumptions made in existing models cause significant limitations in their applicability, and the models themselves are not sufficiently verified. There are no data on the values of thermoelectric and magnetic coefficients for most grades of structural materials, which are necessary for the quantitative assessment of trajectory changes in the conditions of thermoelectric current generation.

The aims of the work are:

- to study the thermo-EMF for steels of different classes
- to establish the dependence of the coefficients of thermo-EMF on temperature
- to estimate the magnetic permeability of materials in the fields which values are determined by the electric currents generating conditions
- to evaluate the shape of the thermo-EMF currents magnetic field in electron beam welding and its correlation with the calculated value obtained by mathematical modeling.

2. Research methods

Using thermocouples, the temperature dependences of thermo-EMF for different materials of various structural classes were obtained. Temperature gradient was provided by placing thermocouple in a high-temperature furnace.

The temperature dependence of thermo-EMF was determined by the integral method (figure 1) on thermocouples of the materials under study (table 1), in which one of the branches was made of chromel.

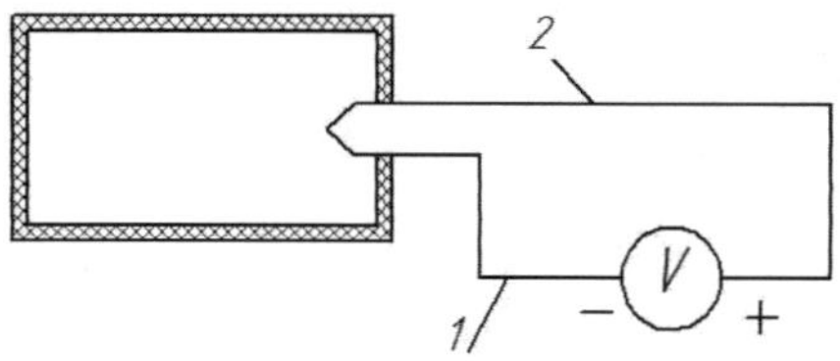

Figure 1. Scheme for integral thermo-EMF study: 1 – chromel, 2 – material under study.

Table 1. The chemical composition of steels (M is martensite, F is ferrite, A is austenite, P is perlite).

No.	Material	Chemical composition, %					Class	Thermo-EMF at 800°C
		C	Cr	Ni	Mo/V	Mn/Si		
1	12Kh18N10T	<0.12	17–19	9–11	–	<2 / <0.8	A	–2.18
2	20Kh13	0.16–0.25	12–14	<0.6	–	<0.6 / <0.7	M+F	7.71
3	09G2S	<0.12	<0.3	<0.2	–/<0.12	1.3–1.7 / 0.5–0.8	P	–0.25

Thermo-EMF was measured with a V7-22 voltmeter when the hot junction was heated in the temperature range of 50–1100°C. Further, the relative Seebeck coefficient for pairs of conductors was calculated by the expression

$$S_r = \frac{dE_r}{dT}, \quad (1)$$

where E_r was a measured relative thermo-EMF.

Then by the expression (2) the absolute Seebeck coefficient at each point was calculated:

$$S_{abs} = S_r - S_{chr}, \quad (2)$$

where S_{chr} was a Seebeck coefficient for chromel.

According to expression 3, the absolute thermo-EMF was calculated for each material from table 1:

$$E_{abs} = \int_{50}^{T} S_x - S_{abs}\,. \qquad (3)$$

The combined PERMAGRAPH® REMAGRAPH®C-750 device, consisting of 2 EF5 fluxmeters, was used for receiving of magnetization curves. Measuring yoke is shown in Figure 2.

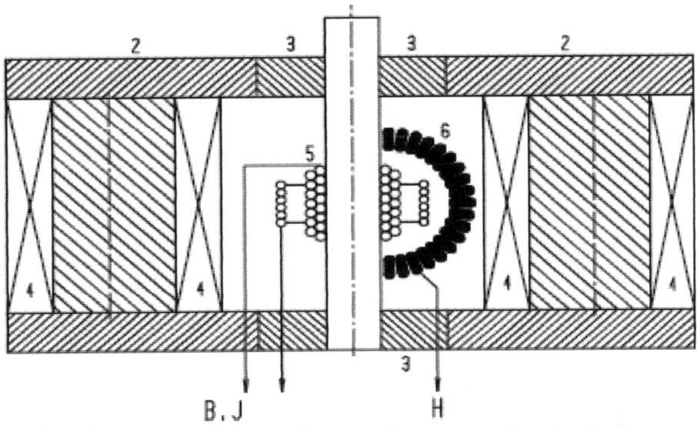

Figure 2. Measuring yoke: 1 — specimen, 2 — soft magnetic steel, 3 — pole pieces, 4 — field generating coils, 5 — J-compensated surrounding coil, 6 — potential coil (for measuring magnetic voltage)

Thus, to receive the magnetization curve, three samples were made for each material under study with a cross-section of 10 x 10 mm (±0.02 mm); the length of each sample was 120 mm.

Before measurements, each of the samples was demagnetized in an alternating magnetic field with a decreasing amplitude. Then, for each of the samples the magnetization curve was received.

The magnetic field strength H in the considered measurement device was determined using a C-shaped coil (Rogowski belt) which was installed directly on the sample surface. With this coil, a magnetic voltage was measured as the potential difference between two points on the sample surface.

The magnetic field strength (in the SI system being measured in amperes) was calculated by dividing the obtained value of the magnetic voltage by the distance between these points:

$$H \approx \frac{U_{ab}^{m}}{l}\,. \qquad (4)$$

In addition to the magnetic field strength, the polarization J, [T] was also determined when the magnetization curve was received. Knowing the polarization J and the magnetic field strength H, the magnetic induction B was determined as

$$B = J + \mu_0 H \qquad (5)$$

The electron-beam welding is characterized by small values of magnetic field induction induced by thermoelectric current, therefore for further analysis they were limited to an induction value not exceeding 100 mT. As in this section the magnetization curve is nonlinear, a high-order polynomial was used to represent the results of measurements.

An improved numerical model of the Seebeck effect in electron beam welding was created on the basis of the obtained data on thermoelectric and magnetic properties.

The model consists of four submodels: thermal problem, electric problem, magnetic and trajectory problem. The thermal problem was solved on the basis of the non-stationary heat equation

$$c(T)\rho(T)\vec{V}\cdot\nabla T = \nabla(\lambda(T)\nabla T) + Q(x,y,z), \qquad (6)$$

where $c(T)$ is the specific heat at constant pressure, J / (kg · K); $\rho(T)$ is the density of the material, kg / m3; $\lambda(T)$ is the thermal conductivity of the material, W / (m · K), $Q(x,y,z)$ is the volumetric heat source, W / m3. The heat source in the EBW (vapor-gas channel) was adopted in the form of a cylinder acting on the entire welded depth. Then, the action of the volume source in equation (6) can be specified as the boundary condition on the cylindrical boundary of the penetration channel with radius r_k in the form of the specific heat flux $q_B(x, y, z)$..

The resulting temperature distribution during welding was used to solve the electrical problem which was formulated as Ohm's law in differential form

$$\vec{j} = \sigma\vec{E} + \vec{j}_e, \tag{7}$$

where $\vec{j}_e = -\sigma S\nabla T$, $\vec{j}$ is the current density vector, $\vec{E}$ is the electric field strength vector, σ is the conductivity of the material, S / m; S - thermoelectric coefficient, V / K.

The magnetic problem was solved regarding the vector potential $\vec{A}$. The magnetic induction distribution $\vec{B}$ was found:

$$\nabla \times \left(\frac{1}{\mu(T)\mu_0}\nabla \times \vec{A}\right) = \vec{j}, \tag{8}$$

$$\vec{B} = \nabla \times \vec{A} \tag{9}$$

In the calculations, the formation of the paramagnetic zone in the region of the penetration channel was taken into account. For this, the dependence of magnetic permeability was set in the form:

$$\mu(T) = \mu_M(1 - h(T - T_c)), \tag{10}$$

where μ_M is the magnetic permeability of the material at room temperature, T_c is the Curie temperature, *h(x)* is the Heaviside function.

And finally, for the obtained distribution of magnetic induction, the trajectory problem of the motion of a charged particle (electron) in a stationary magnetic field was solved :

$$\frac{d^2x}{dz^2} + \sqrt{\frac{2e}{mU}}B_z(x,y,z)\frac{dy}{dz} = \sqrt{\frac{2e}{mU}}B_y(x,y,z) \tag{11}$$

$$\frac{d^2y}{dz^2} - \sqrt{\frac{2e}{mU}}B_z(x,y,z)\frac{dx}{dz} = -\sqrt{\frac{2e}{mU}}B_x(x,y,z), \tag{12}$$

где *x(z)*, *y(z)* – electron trajectories in the XOZ and YOZ planes, respectively, $B_x(x,y,z)$, $B_y(x,y,z)$, $B_z(x,y,z)$ – the projection of the magnetic induction vector on the corresponding coordinate axis; *U* – accelerating voltage in an electron gun.

Due to the complexity of the developed numerical models, their verification should be carried out on several parameters. In addition to the assessment of the electron beam trajectory, the estimation of the magnetic field induction is promising and feasible.

In this case, the control points for measuring magnetic fields should not introduce significant perturbations into magnetic field induced during the EBW; therefore, they should be located outside the body of the welded samples. Figure 3 shows the proposed scheme for measuring magnetic field in the welded samples.

The studies were taken for three pairs of constructional steels – 12Cr18Ni10Ti + 20Cr13 and 09Mn2Si + 20Cr13 – on the samples of 60 mm thick in a form of plates with 60 × 50 × 200 mm dimensions. The welding was carried out on electron beam installation AELTK-344-12 with accelerating voltage of 60 kV. Preliminarily the EBW mode was determined for a weld with point penetration (beam current of 250 mA, welding speed of 15 m/h).

3. Research results

Figure 4 shows the experimental values of the absolute thermo-EMF and the coefficient of thermal EMF for the steels under study.

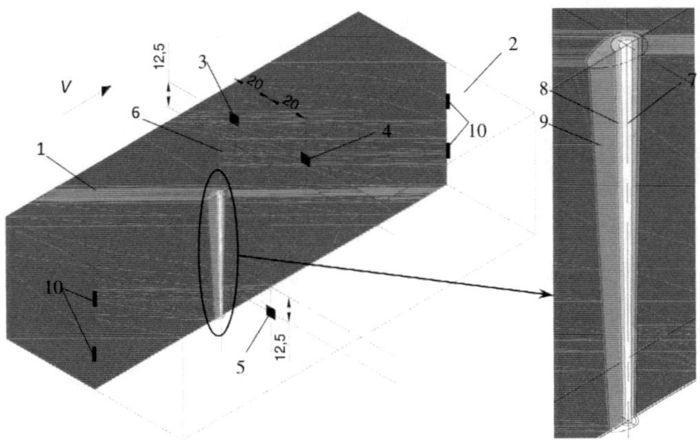

Figure 3. Scheme of magnetic field induction measuring during the EBW of dissimilar materials: 1, 2 – welded samples, 3, 4 – magnetic induction sensors above the samples surface, 5 – magnetic induction sensor in the weld root, 6 – electron beam axis, 7 – electron beam in the penetration channel, 8 – vapor-gas channel, 9 – weld pool, 10 – tack welds.

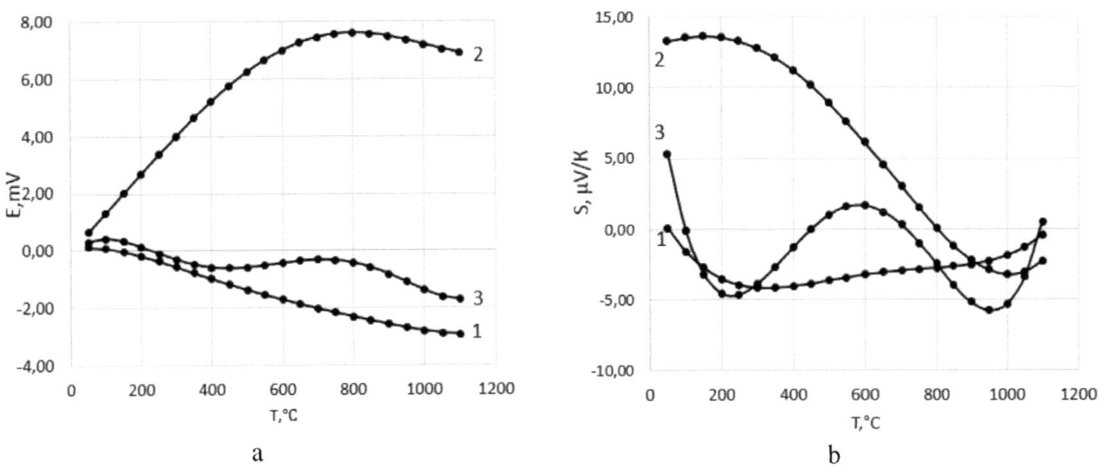

Figure 4. Experimental dependences of absolute integral thermo-EMF and the coefficient of therm-EMF on temperature. 1–3 –number of steels in Table 1

In addition, as a result of the research, experimental tabular and graphical dependences of the magnetic properties for the studied materials were obtained (Figure 5). The obtained results are well approximated by a third-order polynomial. The error in this case ranges from 2% to 9% for various materials (Table 2), which indicates the correctness of the choice of approximating curve.

The results of magnetic field induction measured for such pair of dissimilar steels as 12Cr18Ni10Ti and 20Cr13 are shown in figure 3. The maximum value of magnetic field induction reaches the value of 0.89 mT at the weld root and the value of 0.77 mT above 20Cr13 steel. Electron beam deflection does occur at the weld root of 20Cr13 steel. It is uneven: increasing from 1.8 mm at the beginning of the weld up to 3.8 mm in its middle and then decreasing up to 1.6 mm at the end of the weld (Figure 6). The peak value of the induction at the root is rated at 0.98 mT in the numerical model. It is rated at 0.83 mT above the 20Cr13 steel sample.

IOP Conf. Series: Materials Science and Engineering **759** (2020) 012008 doi:10.1088/1757-899X/759/1/012008

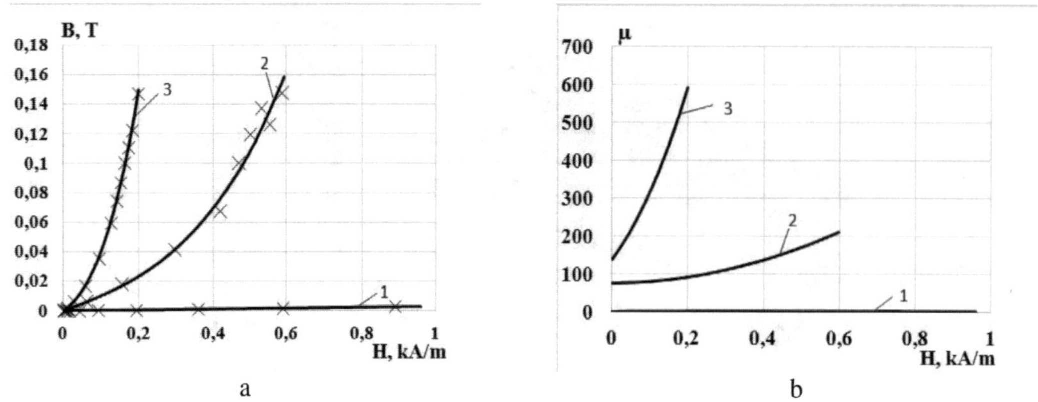

Figure 5. Experimental dependences of the magnetic induction (*a*) and the magnetic permeability of the material (*b*) on the magnetic field strength for the area up to 100 mT. 1–3 –number of steels in Table 1

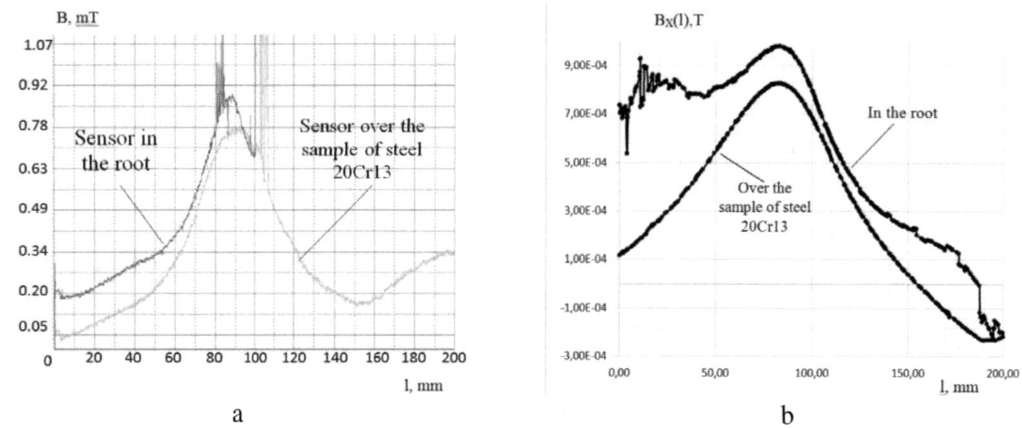

Figure 6. Experimental (a) and numerical (b) magnetic field induction values of thermoelectric currents during the EBW and electron beam deflection in the weld root for 12Cr18Ni10Ti and 20Cr13 steels.

During the EBW of 20Cr13 and 09Mn2Si steel samples the sensors were also installed at a distance of 20 mm from the welded joint and shielded with several layers of aluminum foil. The results of magnetic field induction measuring are shown in figure 5. The maximum values of magnetic field induction were 0.54 mT and 0.47 mT above 20Cr13 and 09Mn2Si steels respectively, having been observed simultaneously. Electron beam deflection at the weld root (Figure 7) is not uneven: it increases from 2.1 mm at the beginning of the weld, reaches a peak of 2.8 mm in the middle of the weld and then decreases towards the end. In the mathematical model the values of the magnetic field induction above the 09Mn2Si steel sample and 20Cr13 steel sample are rated at 0.51 mT and 0.58 mT respectively.

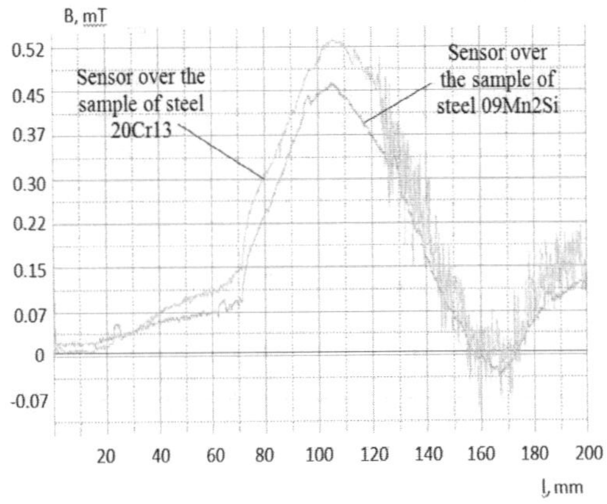

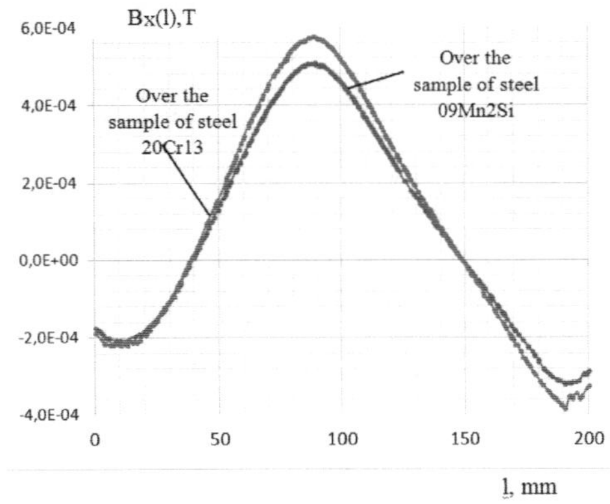

Figure 7. Experimental (a) and numerical (b) Magnetic field induction values of thermoelectric currents during the EBW and electron beam deflection in the weld root for 09Mn2Si and 20Cr13 steels.

4. Conclusion

Thus, the magnetic field induction in the numerical and physical models shows a good correlation. While the EBW modeling of the 12Cr18Ni10Ti and 20Cr13 steels where a significantly greater deflection is observed, the magnitude of the magnetic field also increases compared to the pair of 09Mn2Si and 20Cr13 steels, which indicates the adequate reaction of the numerical model to the welded pair of materials thermo-EMF increasing.

The proposed method of magnetic field measuring at the control points above the welded parts (during the EBW) allows obtaining the data on the magnetic field induction distribution to verify the numerical process models.

Acknowledgments
The research was carried out in National Research University «Moscow Power Engineering Institute» with the financial support of the Russian Science Foundation (project No. 18-19-00652).

References

[1] Wei P S and Lii T W 1990 *J. Heat Transfer* **112(3)** 714-20
[2] Ziolkowski M and Hartmut B 2009 *COMPEL: Int. J. for Computation and Math. in Electrical and Electronic Eng.* **28(1)** 140-53
[3] Dragunov V K and Chepurin M V 2001 *Welding Production* **12** 8–16
[4] Blakeley P I and Sanderson A 1984 *Welding J.* **63(1)** 42–9
[5] Nazarenko O K 1982 *Automatic welding* **1** 33–9
[6] Watanabe K, Shida T, Susuki M *et al.* 1975 *J. Jap. Welding Soc.* **44(2)** 121–7
[7] Dragunov V K, Myakishev Y V, Goncharov A L and Sliva A P 2006 *Welding Int.* **20(10)** 811–5
[8] Goncharov A L 2010 *Welding Production* **4** 12-17
[9] Dragunov V K, Sliva A P, Goncharov A L and Chepurin M V 2014 *Method of dissimilar metallic materials electron-beam welding* (Russian Federation: Patent No 2534183)
[10] Laptenok V D, Druzhinina A A, Murygin A V and SereginYu N 2016 *IOP Conf. Ser.: Mater. Sci. Eng.* **122** 012021

Researches of metal texture after friction stir welding

V E Gozbenko[1, 2], A E Balanovsky[3], S K Kargapol'tsev[2], Yu I Karlina[3], A I Karlina[3]

[1] Angarsk State Technical University, 60 Chaykovskoy Street, Angarsk, 668535, Russia
[2] Irkutsk State Transport University, 15 Chernyshevsky Street, Irkutsk, 664074, Russia
[3] Irkutsk National Research Technical University, 83 Lermontov Street, Irkutsk, 664074, Russia

E-mail: fuco.64@mail.ru

Abstract. The results of studies of the structure of an aluminum alloy by the method of backscattered electron diffraction (EBSD) are presented, which allows determining disorientation angles between grains, present data in the form of an array of angles and sizes, direct and inverse pole figures and many other types of useful information, as well as identify boundaries and sub-boundaries grains. It has been shown that friction stir welding is a process with a predominance of shear; the observed shear texture can indicate the local orientation of the material flow during welding. The three-dimensional material flow detected by these texture orientations shows important changes, especially on the advancing side of the weld.

1. Introduction

In the process of studying the structure and properties of structural materials, we can arbitrarily distinguish two fundamentally different material science methods for describing the microstructure of polycrystals. One of them consists in describing the morphology of crystallites, measuring their characteristic sizes, etc. These research methods can conditionally be called "metallographic". Such experiments, as a rule, are carried out using optical, scanning, and, less commonly, transmission electron microscopy. The purpose of another method is to study the crystal lattice: its primary orientation (texture), internal stresses, etc.; studies are usually carried out using x-ray diffractometry and the methods used can conditionally be called "radiographic". These two ways of integral description of the microstructure for a relatively long time existed in parallel, and with their help significant progress was achieved in studying the evolution of the structure under thermal or mechanical action. Until recently, the tasks of studying texture orientation were solved by almost one single method - using transmission electron microscopy (TEM). A fundamentally new way of studying the microstructure by analyzing the patterns of diffraction of backscattered electrons (EBSD analysis) made it possible to measure its qualitatively new characteristic — disorientation of neighboring crystallites. The method developed in recent years for the automatic analysis of backscattered electron diffraction patterns (EBSD analysis) is also to study the spatial distribution of orientations. In this sense, he is a competitor to TEM. The EBSD method is increasingly used to study the microstructure of crystalline materials after various technological methods of exposure: heat treatment, welding, surfacing, pressure treatment, surface plastic deformation, etc.

Friction stir welding (FSW) is a solid-state bonding process that uses a rotating, non-consumable tool [1,2]. On the one hand, the tool serves as a source of heating of the surrounding material. On the other hand, it mixes the plastic material during rotation and forms a weld. Due to the peculiarities of the process and material flow during FSW, the weld is asymmetric with various zones clearly distinguishable under an optical microscope: mixing zone (MZ), thermo-mechanical influence zone (ZTMI), heat influence zone (HIZ) and base material zone (ZBM) [3]. During friction welding with stirring, a rotating tool deforms the surrounding material mainly due to shear, so the resulting shear strain texture stored in the weld metal can serve as an indicator for the flow of material that occurred around the tool during welding. Texture is a measure of the distribution of the orientation of the crystals in the material and may indicate the type, magnitude and orientation of the deformation. The shear deformation created by the rotating tool is oriented parallel to the surface of the tool and, thus, forms concentric stripes behind the tool, which taper from the top of the weld to the base while reducing the influence of the shoulder and reducing the cross of the tool. From numerous foreign and domestic publications, it is known that finely dispersed grains are formed in the center of the weld (core), the formation of which, according to some researchers [2,3], is the result of dynamic recrystallization [4], collective dynamic polygonization [5], or dynamic return [6]. Taking into account the specifics of the friction stir welding process, it is of practical interest to study the microstructure by analyzing backscattered electron diffraction patterns (EBSD analysis) of crystallite disorientation in the weld zone.

2. Materials and research methods

For research, we selected an industrial thermally not hardenable wrought aluminum – magnesium alloy AMg5 (foreign analog - alloy 5083) in the form of rolled plates with a thickness of 5.0 mm. A friction stir welding seam was obtained at a tool rotation speed of 1400 rpm and a welding speed of 100 mm min-1. The tool geometry used for this weld consisted of a cylindrical pin with a thread and three planes, as well as a rotation arm design. Optical microscopy of a mechanically polished weld was performed in cross section, etched with Keller reagent (2 ml of HF, 3 ml of HCl and 5 ml of HNO3 in 190 ml of water) to reveal the structure of the weld and grains. Electron microscopic studies of the aluminum surface were carried out on the basis of Irkutsk National Research State Technical University using a JIB-4501 JEOL scanning electron microscope with a multipath system equipped with a JIB-4501 electron and ion gun, complete with a nitrogen-free energy dispersive microanalysis system. The analysis of the results of reflected electron diffraction (EBSD analysis) was carried out using the software package for EBSD data processing Channel 5 developed by Oxford Instruments. For these purposes, the studied samples additionally passed the stage of electro polishing. The size of the scanning area for various samples was an area of 100 to 620 μm in size, and the scanning step was from 0.1 to μm. The grain was taken to be a region surrounded by high-angle boundaries, i.e. the value of the limiting angle was set to 15°.

3. Research results and discussion

The alloy structure in the delivery state consisted of grains partially deformed in the rolling direction, while the elongation of the grains is more pronounced for plates of smaller thickness. The average grain size was (25 ± 3) μm.

Figure 1 shows a panoramic survey of the weld, consisting of the area of the core of the weld (1) the mixing zone (MZ), the zone of thermo-mechanical influence (ZTMI) of region 2,3. In region 3, it is clearly seen that the deformation during the friction stir welding proceeds non-uniformly. This allows evaluating the degree of deformation for comparing technological modes and, in particular, to evaluate the geometry of the tool. In Fig. 1, in region 3, regions are observed with layers of the deformable material shifting relative to each other during welding. In the work [7], the adhesive-diffusion nature of the friction stir welding process was shown, when during welding, due to the adhesive-diffusion forces, the material is captured by the tool and moved behind it. In the process of rotation of the tool, the thickness of the layer of the captured material grows to a certain critical mass,

IOP Conf. Series: Materials Science and Engineering **759** (2020) 012009 doi:10.1088/1757-899X/759/1/012009

after which the material comes off the tool and the process repeats. Thus, a layered “onion” structure of the mixing zone of the welded joint is formed. These layers of material move relative to each other, and in places of their contact a local temperature increase due to friction forces is possible. Figure 1 illustrates the bands (region 3) along which the material layers moved during deformation and particles of stable secondary phases deposited on these bands under the action of high temperature.

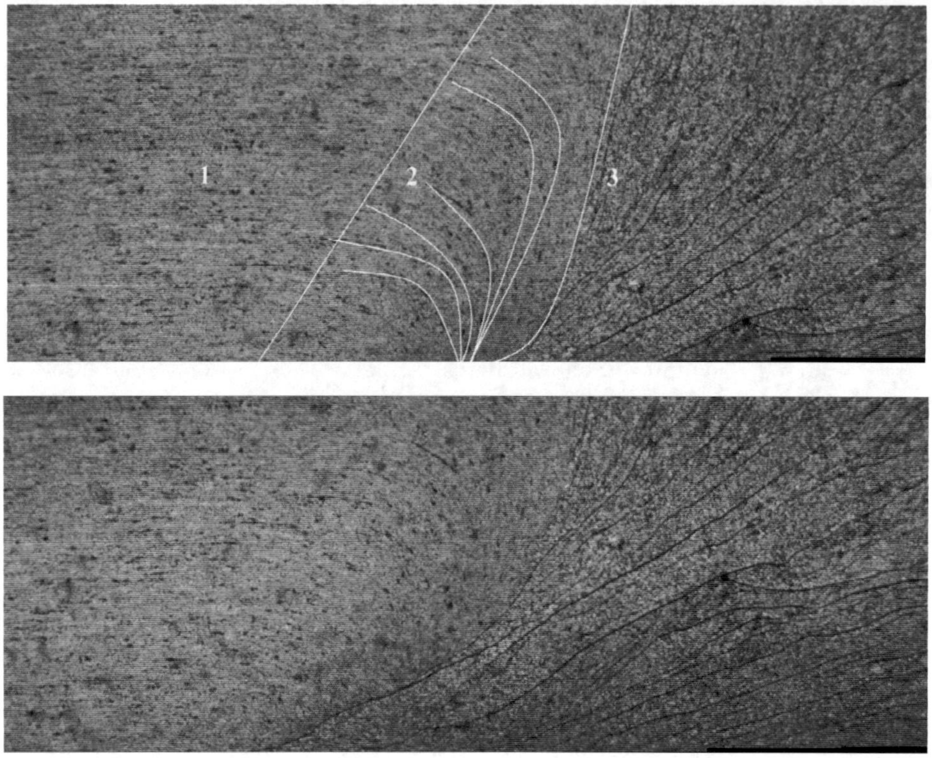

Figure 1. Panoramic shot of the cross section of the weld of the alloy AMG-5.

The mechanism of the formation of a layered structure, as well as during sliding friction, is closely related to the temperature gradient in the surface layer of the deformable material and can be considered on the basis of ideas about the vortex nature of plastic flow. It was shown in the work [7] that such structures are formed during sliding friction as a result of competition between two processes –softening during frictional heating and hardening caused by plastic deformation. A feature of this process is its periodic nature. It is obvious that the heated and plastically deformed weld material was not significantly deformed and any observed deformation texture should reflect the mode, magnitude and orientation of the initial deformation created by the lateral surface of the tool pin. It should be noted that the results of metallographic structural analysis showed that during welding of plates of the AMg5 alloy in the core of the weld (Fig. 1, region 1), a fine-grained structure with the same grain size was formed. To some extent, the grain boundaries were decorated with precipitates of solid particles (oxides).

A detailed analysis of the weld cross section in this regard was carried out by electron backscattering diffraction (EBSD). Scanning a 300 * 600 μm plot is shown in Figure 2.

It should be noted that the study of the processes of texture formation can not be carried out separately from the analysis of the general change in the structure of the material with appropriate thermo-mechanical processing. Thus, the formation of deformation textures depends on the mechanism of plastic deformation. Certain information can be obtained by analyzing the distribution of the preferred orientations of the material in Figure 2. For visualization of orientations of structural elements (grains), orientation maps were constructed. Such maps can be built both in the space of Euler angles and inverse pole figures Fig. 2, b. In Euler space, each point in the scanning area is assigned a specific orientation, characterized by three Euler angles, which give an unambiguous position of the unit cell in space. An alternative to the space of Euler angles is the space of inverse pole figures. In this space, each scanning point is assigned Miller indices (Fig. 2, c.), which also give an unambiguous position of the unit cell in space. At the vertices of the triangle are Miller indices (111), (101), (001). Thus, if the point (and subsequently the whole grain) on the surface of the studied material has an orientation of (111), then it is colored blue; if (001), then - in red; if (101), then in green. In the case when the point has a different orientation from the data, it is painted in the color corresponding to the legend. Figure 2 shows orientation maps for the samples under study. Figure 2a shows a contrasting image of the grain structure. The analysis of the presented data shows how uniformly the orientations in the material under study are distributed. At first glance, a chaotic distribution of grain orientation is characteristic. At the same time, after friction stir welding, regions of grain orientations of Fig. 2a and grain boundaries that gravitate toward orientation (101), which are colored green (Fig. 2, c.).

This fact is easy to explain: in bcc metals, the main slip plane is (101); therefore, the observed increase in the area of green areas on orientation maps is deformational. For more reliable discussions about the deformation mechanisms in terms of "orientations", it is necessary to conduct a comprehensive analysis of orientation maps and pole figures (both direct and inverse). As follows from the results of Fig. 2, the average grain size decreased by 4-5 times compared to the initial grain structure of the alloy. The formation of such a structure can be explained by the intense thermo-mechanical action during friction of a rotating tool and the implementation of dynamic recrystallization under these conditions. It should be noted that there is no consensus among researchers on this issue. One of the important evidences of the dynamic recrystallization process is the determination of the disorientation angle of the formed grain boundaries, which gives a reliable idea of what type of these boundaries are - large-angle or small-angle. It is generally accepted that the high-angle boundaries characteristic of recrystallized grains includes those whose disorientation angle exceeds 15°. The backscattered electron diffraction (EBSD) method allows determining orientation angles with an accuracy of ~ 2°.

According to the data obtained, the average grain size in this zone (core) of the weld was 4.3 μm, which is in good agreement with the above results of optical metallography. From Fig. 2, d (highlighted in black) it follows that the fraction of boundaries with small disorientation angles (small-angle boundaries) amounted to about 1 % of the total number of boundaries, which allows concluding that this structure is formed mainly by equiaxed grains with high-angle boundaries (Fig. 2, g red color). The average grain size correlates well with the data obtained by optical microscopy in this zone of the weld. A distinctive feature of the plastic deformation of metals is that the deformation begins only when the external stress in the crystal reaches a critical shear stress or yield strength in the plane and in the direction in which it occurs, which corresponds to the law of critical shear stress (Schmid's law).

At the same time, out of all possible equal systems of directions and sliding planes, the first system to start functioning is the one in which the external voltage component reaches its maximum value first. With certain orientations of the crystal, the shift can begin simultaneously in several slip systems. The backscattered electron diffraction method (EBSD) allows realizing the possibility of constructing micro-stress distribution maps on the surface of the test sample. Such maps are constructed as follows: the program measures the maximum disorientation between two arbitrary points within the same grain. Then, using the mathematical "weighting" procedure, each point is assigned a specific weight value according to the maximum disorientation measured in the previous step. Then, in automatic mode, builds a smoothed map of the distribution of micro-stresses according to the average grain size of the structure of the studied material. Figure 3 shows the results of constructing orientation maps of structural elements (grains) over an area of 50 * 50 μm. The first picture shows the orientation of the grains along the (111) plane in color, it is painted in blue; (001) - in red; (101) green. Blue color and

IOP Conf. Series: Materials Science and Engineering **759** (2020) 012009 doi:10.1088/1757-899X/759/1/012009

its combinations have a large area, and the slip plane (101) green color has a smaller area. In the second picture, special twinning borders are plotted. The third picture shows microstresses in the area of grain boundaries. The maximum microstresses are fixed in fine grains.

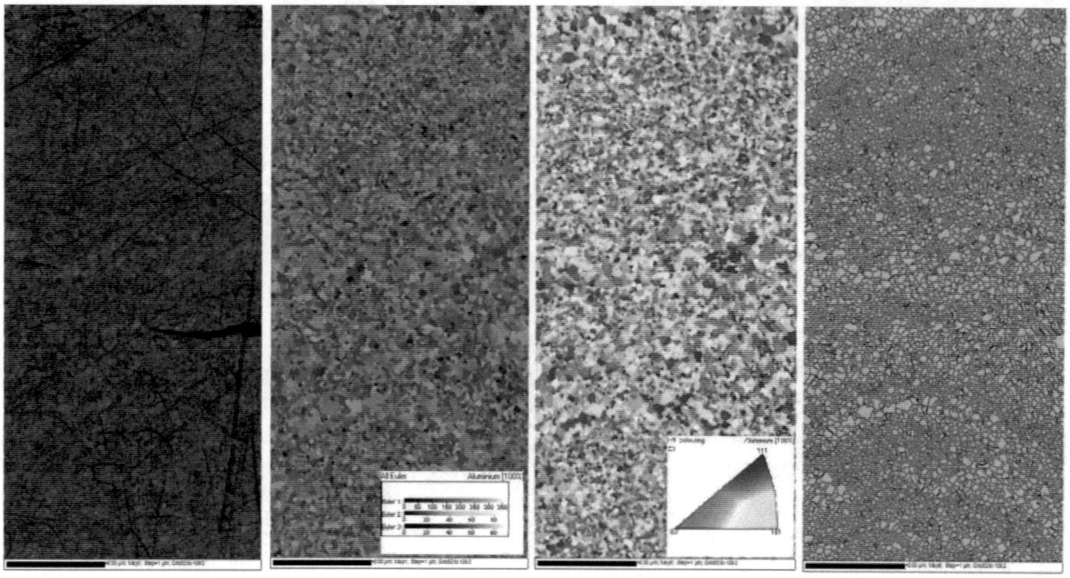

(a) (b) (c) (d)

Figure 2. Results of studying the grain structure of the AMG-5 alloy by electron backscattering diffraction (EBSD).

It should be noted that at present, with the traditional method of aluminum welding [8,9], the mechanism of releasing the weld from the oxide film is not yet fully understood. The role of cathode spots is constantly being reviewed. Many authors base their hypotheses on the basis of evaporation and cathodic cleaning mechanisms. In this regard, in friction stir welding, the role of the oxide film in the structure of the weld and the mechanism for its removal are not taken into account at all. It is quite obvious that at a metal temperature during the friction welding with stirring of 0.8 T_m, the oxide film is not removed, but mixed (moved) with the weld metal. This should affect the mechanical properties of the weld.

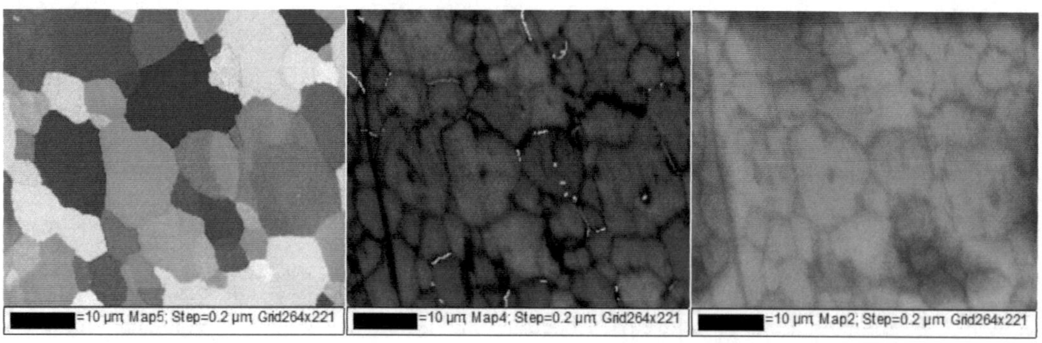

Figure 3. Results of the study of the grain structure of AMG-5 alloy by electron backscattering diffraction (EBSD).

In this regard, we must take this factor into account when welding, but for this it is necessary to find in micro-structural studies the areas where this film is concentrated (located). Under certain

IOP Conf. Series: Materials Science and Engineering **759** (2020) 012009 doi:10.1088/1757-899X/759/1/012009

conditions of sample preparation by electron backscattering diffraction (EBSD), it is possible to study such areas.

4. Conclusion

In conclusion, the following should be noted: the electron backscattering diffraction method (EBSD) can be used to study aluminum alloys and give a complete overview of the structure features at the micro level, and in combination with texture analysis, it allows establishing the physical mechanisms of structure formation during deformation. At present, a huge number of approaches have been developed to the task of attesting the structural state of deformed materials; however, the choice of various tools affects the results obtained by the researcher. Therefore, the correspondence to the task posed by the researcher and the means for its solution is important.

References

[1] Thomas W M, Nicholas E D, Needham J C 1991 Pat. Appl. no. 9125978.8 GB Friction stir butt welding – International patent application No. PCT/GB92/002203GB
[2] Frolov V A, Ivanyukhin A N, Sabantsev A N 2008 *Welding production* **10** pp 12-19
[3] Kotlyshev R R, Chularis A A, Ludmirsky Yu G 2010 *Welding and diagnostics* **4** pp 31-35
[4] Fonda R W, Knipling K, Bingert F 2007 *Scripta Materiala* Vol. 58 pp 343-348
[5] Frolov V A, Konkevich V Yu, Predko P Yu, Belotserkovets V V 2013 *Welding production* **3** pp 21-26
[6] Dobatkin S V, Zakharov V V, Perevezentsev V N 2010 *Technology of light alloys* **1** pp 74-84
[7] Rubtsov V E, Kolubaev A V 2004 *Letters in ZhTF* Vol. 74 **11** pp 63-69
[8] Balanovskii A E 2018 *High Temperature* **56(1)** 3-13
[9] Balanovskii A E 2018 *High Temperature* **56(3)** 329-337

IOP Conf. Series: Materials Science and Engineering **759** (2020) 012010 doi:10.1088/1757-899X/759/1/012010

Research of mechanical properties and chemical composition of samples made of aluminium alloy AMg6 and obtained by electron beam additive manufacturing with wire feeding

A V Gudenko[1,3], G V Grigoriev[2], A P Sliva[1], R Yu Agafonov[2], A N Kasitsyn[2]

[1]National Research University "Moscow Power Engineering Institute", 14, Krasnokazarmennaya st., Moscow, 111250, Russia
[2]Corp. "Russian Space Systems", 53, Aviamotornaya st., Moscow, 111250, Russia
[3]E-mail: alexsguddy@gmail.com

Abstract. This paper is devoted to the definition of mechanical properties, porosity and chemical composition of the samples made of wrought aluminium alloy AMg6 and obtained by electron beam additive manufacturing with wire feeding. It has been established that ultimate strength of the deposited metal is 90 % of the rolled metal. Chemical composition of the samples has not differed significantly from wire. According to rough estimations porosity has not exceeded 0.055%.

1. Introduction
One of the most perspective technologies in the field of additive manufacturing is electron beam additive manufacturing with wire feeding. In addition to advantages of the wire feed additive technologies [1] the process is carried out in a vacuum chamber that has a positive effect on quality of deposited material. In addition, the use of electron beam as a power source and vacuum chamber makes it possible to produce parts form wide range of materials such as titanium alloys [2] actively interacting with atmospheric gases, nickel alloys [3], stainless steels [4] and fusible aluminium alloys [5]. However, there is a lack of attention to investigation of additive manufacturing with the use of an electron beam and aluminium wire.

Electron beam additive manufacturing with aluminium wire could be used to produce large-sized details with complicated shape of walls because additive technologies lead to increase in the rate of the use of materials. Moreover, this technology could be applied to restoration a shape of parts in a repair operation.

An example of such parts is an aluminium case for electronic hardware (figure 1) that is used in aerospace industry where the main task is to decrease mass and dimensional characteristics without reducing mechanical properties.

The purpose of this paper is to study the possibility to use electron beam additive manufacturing with wire feeding for producing parts of aluminium alloy AMg6 and restoring their shape and dimensions in a repair operation.

2. Research methods
Two specimens made of aluminum alloy AMg6 in the form of walls were deposited to examine chemical composition, porosity and mechanical properties.

An AMg6 wire with the diameter of 1.6 mm was used. Two AMg6 blocks with dimensions of 120x28x20 were used as a substrate.

Chemical composition of AMg6 alloy is shown in Table 1.

Table 1. Chemical composition of AMg6 alloy in % according to GOST 4784-97.

Fe	Si	Mn	Ti	Al	Cu	Be	Mg	Zn	Other
< 0.4	< 0.4	0.5-0.8	0.02-0.1	91.1-93.68	< 0.1	0.0002-0.005	5.8 - 6.8	< 0.2	< 0.1

The deposition was carried out using an electron beam machine with an accelerating voltage of 60 kV that was equipped with an automatic wire feeder. The wire feed rate was 2 m/min and the deposition rate was 60 m/h. Before the deposition of the first layer the trajectory was cleaned by an electron beam. The beam power was 360 W during cleaning pass.

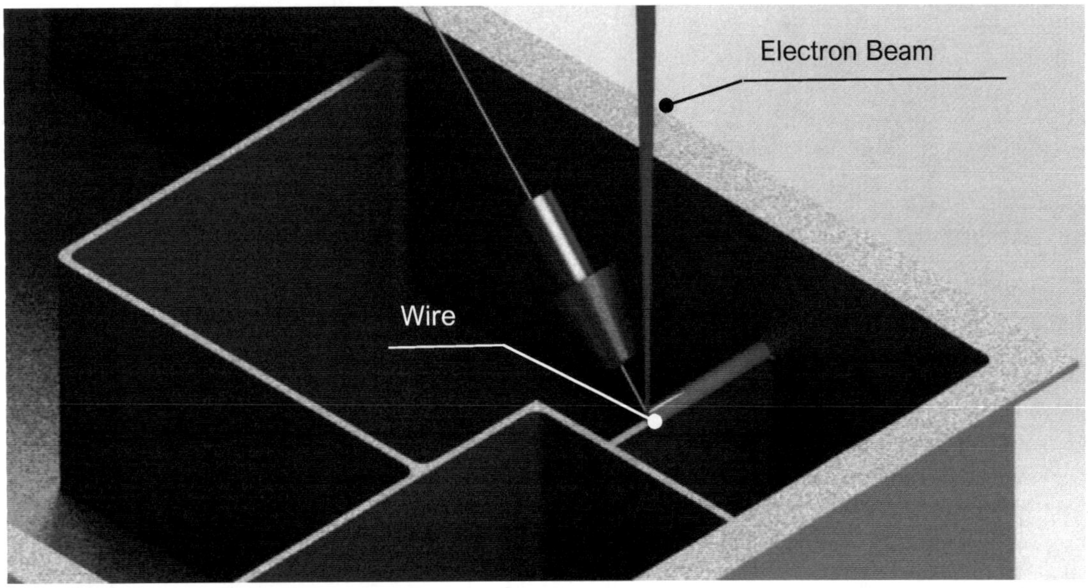

Figure 1. The reconstruction scheme of the reconstruction of the original size of radioelectronic equipment case walls.

The deposition modes were chosen in accordance with the calculation method [6] and were adjusted experimentally (Table 2). Power decrease from layer-to-layer is necessary for obtaining a wall with constant thickness since an increase of layers number leads to deteriorating of heat transfer conditions as a result of temperature growth of the substrate and upper layer during deposition process. The change of the deposition conditions, in its turn, leads to the growth of melt pool temperature and the reduction of surface tension leading to width increase and height decrease of the deposited layer. However, heat transfer stabilizes at the certain layer and further decrease of electron beam power can lead to interruption of the deposition process.

Table 2. Adjustment of electron beam power at each layer.

	Beam power, W	
Layer number	Sample 1	Sample 2
1	1140	1200
2	1020	960
3	780	840
4 - 12	720	780

An oscillation of type "3 concentric circles" was applied. The diameters of the circles were 3, 2 and 1 mm. The oscillation frequency was 1000 Hz. It has previously been established that the use of this type of oscillation leads to the more stable formation compared to the "circle" type. This type of oscillation results in the more equal distribution of power density in the treatment area. It is necessary to redistribute power density more equally because of intense evaporation of aluminum alloy at the high densities which results in instability. The deposited specimen is shown in Figure 2.

The specimens were mechanically processed and further radiography, chemical and mechanical tests were carried out. Machined specimens are shown in Figure 3. Radiography tests were done to detect pores and other defects and to assess their dimensions. Rough estimation of porosity was carried out by calculating pores volume that were visible in the X-ray picture and dividing this volume by the volume of metal depicted.

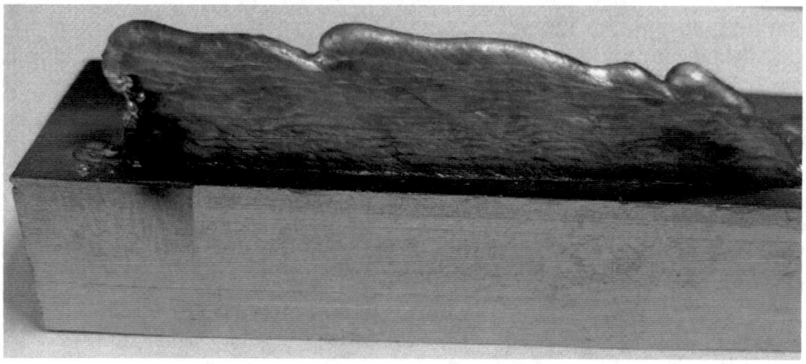

Figure 2. Deposited specimen

The specimens for the tensile tests had dimensions of 55x8x1.5 mm (flat specimens were made in accordance with GOST 1497-84). The deforming velocity was 5 mm/min. Ultimate strength values of deposited metal have been established.

Chemical composition tests (GOST 7727-81) were performed with the use of emission spectrometer Foundry Master to assess chemical composition changes during deposition process.

Figure 3. Specimens after milling

3. Results and discussions

Based on radiography (figure 4) it has been established that pores have been distributed evenly by volume. The size of pores has not exceeded 100 μm for each specimen. Some pores with size of about 200-250 mm could be observed. According to rough estimation porosity has not exceeded 0.055%. Also, it could be noted that fine pores are concentrated in the bottom area of the specimens.

Porosity in the specimens has been connected to preparation quality of the wire surface. The surface layer of the wire adsorbs water steam that evaporates and makes cavities after crystallization. It could be possible to reduce the number of pores by increasing the time during which metal stays melted, that is, by decreasing deposition rate. On the other hand, increasing the time during which metal stays melted may lead to the evaporation of some elements which have high vapor pressure that could cause negative impact on mechanical properties of a part. Thus, it is safe to say that it is necessary to adjust modes in such a way that it does not lead to porosity and significant evaporation of elements e.g. magnesium. It is also necessary to prepare wire surface carefully.

The ultimate strength of the specimens has been established based on tensile tests. The results of the test and the ultimate strength of cast, wrought, and rolled materials are shown in Table 3.

The ultimate strength of deposited material has exceeded by 40 % this parameter of cast material. Such exceedance is related to the high speed of crystallization that leads to refinement of grains and to the increase of mechanical properties.

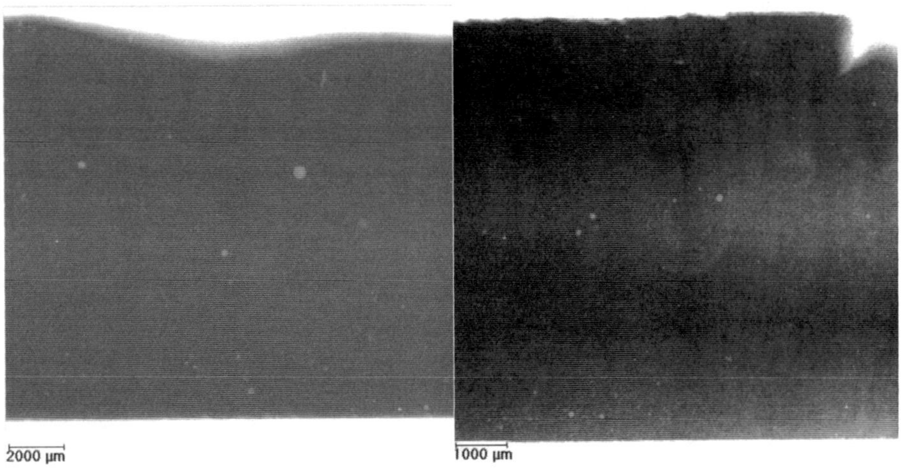

Figure 4. Radiography pictures of the specimens 1 (left) and 2 (right).

The ultimate strength value of deposited metal has been about 90% of rolled metal. The lower value could be explained by the fact that the deposited metal is not strain hardened. Moreover, there are pores in the deposited metal that decreases its mechanical properties.

It should be considered that mechanical properties were investigated in the same direction as the direction of deposition. In other directions properties could differ significantly [7].

Table 3. Mechanical properties of deposited metals and their comparison with cast, wrought, and rolled metal.

		UTS, MPa
Additive manufacturing	Specimen 1	303
	Specimen 2	294
AMg6, cast [8]		210
AMg6, wrought [9]		330
Amg6, rolled [10]		340

Studies of chemical compositions have been carried out in several areas of both specimens (figure 5).

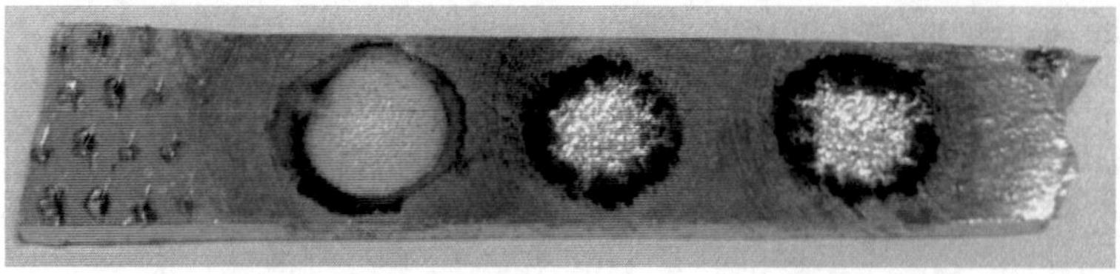

Figure 5. Areas of chemical composition testing.

The chemical composition of deposited specimens has not differed significantly from the chemical composition of the wire (table 4). High feed rate and not significant overheating of melt pool do not lead to much evaporation of magnesium. It could be said that there is no significant evaporation of magnesium during deposition of AMg6 alloy that could lead to loss of strength of material.

Table 4. Chemical composition of deposited metal and wire, %

Element	Wire AMg6, %	Deposited specimens (average), %
Al	balanced	balanced
Si	0.124	0.116
Fe	0.312	0.224
Cu	0.081	0.057
Mn	0.690	0.650
Mg	**6.214**	**5.980**
Zn	0.034	0.029
Ni	0.004	0.003
Ti	0.145	0.103
Be	0.002	0.001
Pb	0.018	0.010

4. Conclusions

Based on the research on deposited specimens of aluminum wrought alloy AMg6 it has been established that the size of pores has not exceeded 100 μm in the majority cases. However, some pores with the diameter of 200-250 μm have been observed. The size of pores largely depends on the quality of wire surface preparation and on time during which metal stays at high temperatures.

Chemical composition has not changed significantly as a result of deposition. There is not much evaporation of elements with high vapor pressure such as magnesium that leads to loss of strength.

The ultimate strength of specimens studied in the direction of deposition has been 40% more than ultimate strength of cast metal and has been about 90% of rolled metal. High mechanical properties are related to high deposition rate that leads to grain refinement.

Thus, electron Beam additive manufacturing with wire feeding with the use of alloys such as AMg6 could be used to manufacture parts with complicated shape and with the mechanical properties that are higher than the properties of cast metal as long as it is cost-effective.

Acknowledgements

The research was carried out in National Research University "Moscow Power Engineering Institute" with the financial support of the Russian Science Foundation (project No. 17-79-20015).

5. References

[1] Ding D, Pan Z, Cuiuri D, Li H 2015. *Int J Adv Manuf Technol* **81**(1–4):465–81
[2] Gurianov D A et al. 2019 *IOP Conf. Series: Material Science and Engineering* **597** 012042
[3] Xu X, Ding J, Ganguly S, Williams S 2019 *J of Materials Processing Tech.* **265** 201–209
[4] Tarasov S Yu et al 2018 *J. of Alloys and Compounds* **803** 364-70
[5] Brice C et al 2015 *Materials Science and Engineering A* **648** 9-14

[6] Gudenko A V et al 2018 *Svarochnoe proizvodstvo* **8** 12-19 (In Russian)
[7] Bird R K, Atherton T 2010 *NASA Tech Memorandum* (TM-2010-216719)
[8] Chuvil'deev V N et al 2008, Mechanical properties of microcrystalline AMg6 aluminum alloy *Vestnik NNGU* **4** 35-42 (In Russian)
[9] Fridlander I N 2001 *Non-ferrous metals and alloys. Composite metal materials.* (Moscow: Mashinostroenie) (In Russian)
[10] Miklyaev P G 1994 *Mechanical properties of lightweight alloys in conditions of temperatures and forming processes velocities* (Moscow: Metallurgy) (In Russian)

IOP Conf. Series: Materials Science and Engineering **759** (2020) 012011 doi:10.1088/1757-899X/759/1/012011

Possibility of obtaining complex form details using additive surface technology

S K Kargapol'tsev[1], **A E Balanovsky**[2], **V E Gozbenko**[1, 3], **Yu I Karlina**[2], **A I Karlina**[2]

[1] Irkutsk State Transport University, 15 Chernyshevsky Street, Irkutsk, 664074, Russia
[2] Irkutsk National Research Technical University, 83 Lermontov Street, Irkutsk, 664074, Russia
[3] Angarsk State Technical University, 60 Chaykovskoy Street, Angarsk, 668535, Russia

E-mail: fuco.64@mail.ru

Abstract. The authors consider the possibilities of using welding in protective gasses with a melting electrode for the manufacture of parts of complex shape. The results of surfacing parts such as square and cylinder are presented. In the course of the research, the technological features associated with the parameters of the welding arc and melting of the surfacing wire were established. Arc current has the greatest effect on the appearance of the molding. The wire feed speed affects the height of the deposited roller. It is shown that the stability of the formation of deposited rollers depends on the transfer modes of the metal, which depend on the conditions of supply and input of wire into the melt pool. The additive technology of metal deposition allows one obtaining parts with a surface different from the base metal, high wear resistance at normal and elevated temperatures, and corrosion resistance. Surfacing can be performed both in the manufacture of new parts and in repair and restoration work, significantly extending the life of parts and assemblies, thereby ensuring a high economic effect.

1. Introduction

Additive technology. (AT) today are a modern and promising method for the production of parts and products of complex configuration [1-12]. They are based not on the "subtraction" of the material, but on its addition, layer-by-layer building up of the material in order to create an object based on an electronic three-dimensional model. These technologies make it possible to use practically the same amount of material during manufacturing that will subsequently remain in the finished part. Due to this approach, the utilization rate of the material increases and it becomes possible to produce parts that cannot be obtained by casting or machining. AT is universal, flexible, easily customizable and therefore can be used in most sectors of industrial production. The materials for the manufacture of various parts of machine parts and units can be of various types. These include metallic, ceramic, and polymeric materials, as well as combinations of composite, hybrid, or functionally gradient materials (FGMs). The most interesting and effective technologies for the manufacture of metal products by the method of layer-by-layer application of material as applied to industry are the following: WAAM (GMAW, PAW, CMT), DMD, LBDMD, EBAM. WAAM - Wire Arc Additive Manufacturing) is one

IOP Conf. Series: Materials Science and Engineering **759** (2020) 012011 doi:10.1088/1757-899X/759/1/012011

of the additive technologies that use wire as a surfacing material [7-12]. Various research groups are working on improving such technologies of wire-arc additive production as metal electrode welding in a gas medium (GMAW), gas-electric welding with a tungsten electrode (GTAW) and plasma arc welding (PAW) [12-19]. GMAW is a type of welding, otherwise called MIG / MAG welding (arc welding with a consumable electrode in an inert gas environment), which involves the automatic supply of a continuous solid melting electrode in a protective external gas atmosphere.

2. Materials and research methods

The energy source selected for this AT-process is an electric arc with a floating electrode (GTAW), and not a laser or electronic beam, due to a combination of advantages, including low cost, no need for a vacuum chamber (compared to electron beam), the ability to economically apply the process for a very large working volume and a much higher deposition rate. The GTAW process can be combined with industrial welding robots and multi-sensor control systems to achieve high accuracy in arc positioning. The experimental setup includes an industrial current source, a wire feeder, a carbon dioxide cylinder, a work table and a working chamber. In the manufacture of parts of complex shape, the main geometric parameters that must be controlled are the effective wall width and layer height. These sizes are set from the initial parameters of the weld pool and then the rollers are superimposed on each other. Figure 1 shows a fragment of the formation of a cylindrical part.

Figure 2 shows a typical surface profile of a complex part in the form of a square made on the basis of GMAW. Part of the weld square was machined. As it is shown in Figure 3, the uneven appearance of the deposited surface is observed noticeably in the adjacent layers, due to the layer height of more than 1 mm, during the deposition, many deposited layers. This phenomenon is known as the step effect, which inevitably arises in additive manufacturing. Due to the greater layer height, this effect is more predominant in WAAM compared to the effect of selective laser melting (SLM).

In general, the surface quality of the deposited parts has a significant effect on the amount of processed metals in the process of subsequent finishing. A fully dense material is necessary for the manufacture of most parts in order to achieve optimum mechanical strength to meet or exceed service requirements. Full density is achieved by optimizing both process parameters and the trajectory of the welding head to maintain a continuous melt pool. Figure 3 shows layers consisting of a cellular microstructure. Continuity from layer to layer is maintained by melting and back-melting into previous layers when a new metal layer is added. The depth of reverse melting varies from a fraction of the previous layer to repeated melting through several layers, depending on the selected process parameters.

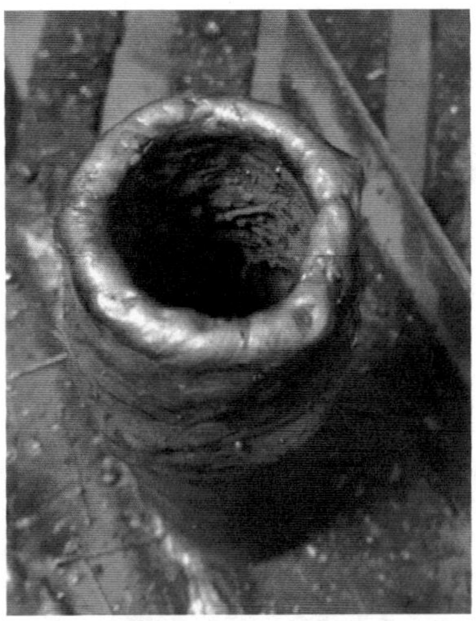

Figure 1. Additive technology of semi-automatic CO_2 deposition technology to produce cylinder-shaped parts.

Figure 2. Additive technology of semi-automatic deposition technology in CO_2 to obtain a square part.

On the whole, during the experiments, a good formation of deposited layers was established. The results of mechanical processing showed the absence of pores and cracks in the deposited wall of the square of Figure 3. No fusion between the layers is not fixed. At the same time, it is necessary to note the presence of defects during surfacing (Fig. 3).

With semi-automatic deposition of metal on the surface of a product with the participation of a welder, the following technological difficulties arise:

- during surfacing it is often necessary to take breaks;
- with increasing height of the deposited metal, the shape of the circle begins to decrease;
- due to severe overheating, the metal begins to spread, thereby increasing the surface area of the part;
- it is possible to change the specified form of surfacing.

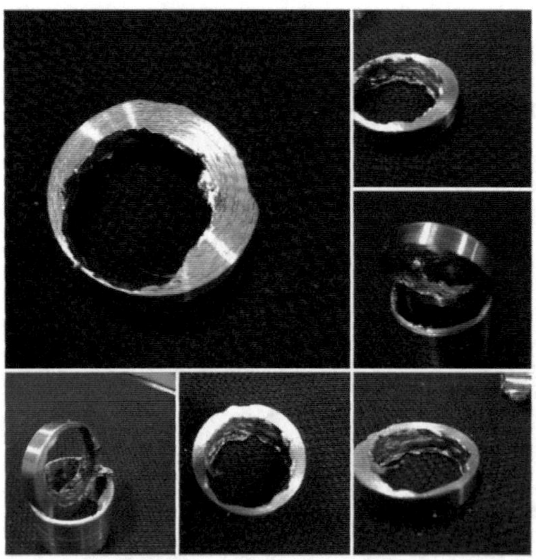

Figure 3. Surfacing defects.

The change in the shape of the circle (see Fig. 3) is due to the fact that, for the initial base, it is necessary to lay the rollers of a larger area. Changing the shape of the deposition depends on the welder. the latter, for subjective reasons, cannot surface up with constant speed.

Under suboptimal conditions, the formation of porosity of the microstructure is possible, which occurs mainly as a result of gas evolution during solidification and the absence of melting between layers or adjacent passages of the molten bath. Increasing the power of the welding arc, reducing the speed of movement, or using thinner layers can help merge and reduce or eliminate this type of voids. Based on the experimental results, it was found that the arc current has the greatest influence on the appearance of the molding.

3. Conclusion

In the course of the research, the technological features associated with the parameters of the welding arc and melting of the surfacing wire were established. Arc current has the greatest effect on the appearance of the molding. The wire feed speed affects the height of the deposited roller. It is shown that the stability of the formation of deposited rollers depends on the transfer modes of the metal, which depend on the conditions of supply and input of wire into the melt pool.

he additive technology of metal deposition allows one obtaining parts with a surface different from the base metal, high wear resistance at normal and elevated temperatures, and corrosion resistance. Surfacing can be performed both in the manufacture of new parts and in repair and restoration work, significantly extending the life of parts and assemblies, thereby ensuring a high economic effect.

References

[1] Herranz S, Campa FJ, Lopez de Lacalle LN, Rivero A, Lamikiz A, Ukar E, Sanchez JA, Bravo U 2005 *The milling of airframe components with low rigidity: a general approach to avoid static and dynamic problems. Proc Inst Mech Eng, B: J Eng* Manuf pp 789-801

[2] Ospennikova O G, Evgenov A G, Nerush S V, Afanasyev-Khodykin A N 2012 *Vestnik UGATU* **5 (50)**

[3] Nerush S V, Evgenov A G, Ermolaev A S, Rogalev A M 2013 *Problems of Materials Science* **4 (76)**

[4] Evgenov A G, Nerush S V, Vasilenko S A 2014 *Transactions of VIAM* **5**

[5] Lewis G, Cremers D, Cotton J, Milewski J, Preston D. *Laboratory directed research and development program annual reports for FY92, FY93, FY94. The rapid formation of unique structural components by fusing airborne powders in a laser beam* (Los Alamos National Laboratory, Los Alamos New Mexico) 97545

[6] Lewis G K, Nemec R B, Milewski J O, Thoma D L, Barbe M R, Cremers D A 1994 *Directed light fabrication* (Proceedings of the ICALEO '94. Laser Institute of America, Orlando, Florida) 17

[7] Smugeresky J E, Keicher D M, Romero J A, Griffith M L, Harwell L D 1997 *Proceedings of the World Congress on Powder Metallurgy and Particulate Materials. Chicago, IL*

[8] Griffith M L, Keicher D M, Atwood C L et al. 1996 *Free form fabrication of metallic components using laser engineered net shaping (LENS). Proceedings of the Solid Freeform Fabrication Sym-posium. Austin* 125

[9] Oskolkov A A, Matveev E V, Bezukladnikov I I, Trushnikov D N, Krotova E L 2018 *Bulletin of PNIPU. Engineering, material management* Vol. 20 **3** pp 90-105

[10] Jun Xiong, Guangjun Zhang, Weihua Zhang 2015 *The International Journal of Advanced Manufacturing Technology* Vol. 80, Issue 9-12, pp 1767-1776

[11] Ding D [et al.] 2015 *International Journal of Advanced Manufacturing Technology* Vol. 811 **4** pp 465-481

[12] Bonaccorso F, Cantelli L, Muscato G 2011 *IFACProceedings Volumes* Vol. 44 pp 11636-11641

[13] Xiong X [et al.] 2009 *J Mater Process Technol* **209** pp 124-130

[14] Wang H, Jiang W, Valant M, Kovacevic R 2003 *Proceedings of the Institution of Mechanical Engineers* **217** pp 1641-1650
[15] Martina F [et al.] 2012 *F. J. Mater. Process. Technol.* Vol. 212 № 6 pp 1377–1386
[16] Williams S W [et al.] 2015 *Materials Science and Technology* Vol. 32 iss. 7 pp 641–647
[17] Cong B, Ding J, Williams S W 2015 *J. Adv. Manuf. Technol.* № 76. pp 1593–1606
[18] Cheng B, Chou K 2015 *Computer-Aided Design.* № 69 pp 102–111
[19] Baufeld B. 2016 *Elektronika ir Elektrotechnika* Vol. 22 № 5–6 pp 7–13

Challenges in path planning of high energy density beams for additive manufacturing

Karunakaran K P [1,6], Seema Negi [1], Arun Nambolan[1], Ashik Patel [1], Yogesh Patil[1], Milan Pandya[1], Sajan Kapil [2], Dmitriy Trushnikov[3], Michael Sealy[4] and Alain Bernard[5]

[1] Indian Institute of Technology Bombay, Main Gate Rd, IIT Area, Powai, Mumbai, Maharashtra 400076, India
[2] Indian Institute of Technology Guwahati, Surjyamukhi Road, North, Amingaon, Guwahati, Assam 781039, India
[3] Perm National Research Polytechnic University, Komsomol'skiy Prospekt, 29, Perm, Perm Krai, 614990, Russia
[4] University of Nebraska-Lincoln, 1400 R St, Lincoln, NE 68588, United States
[5] Ecole Centrale de Nantes, 1 Rue de la Noë, 44300 Nantes, France
[6] E-mail: karuna@iitb.ac.in

Abstract. As there are no cutting forces in High Energy Density (HED) beams like lasers and Electron Beam (EB), their speeds are limited only by their positioning systems. On the other hand, as the entire matrix of the 3D printed part has to be addressed by the thin beam in multiple passes in multiple layers, they have to travel several kilometers in tiny motions. Therefore, the acceleration of the motion system becomes the limiting factor than velocity or precision. The authors have proposed an area-filling strategy for EB to fill the layer with optimal squares to exploit analog and hardware computing. 3D printing requires uniform intensity slanged as flat hat shape whereas the default is Gaussian. The authors have proposed an optimal algorithm that takes into account the maximum velocity and acceleration for achieving a flat hat without any compromise on productivity.

1. Introduction

In milling, the cutter and workpiece are always in contact to cause separation of material using forces (Figure 1a). So, friction at the tool-workpiece interface limits the traversal speeds. Usually, these are not greater than 3 m/min (0.05 m/s). Despite these low speeds, productivity is high as the milling cutters are large. On the other hand, *High Energy Density (HED)* beams like laser and *Electron Beam (EB)* do material removal by energy transfer [1-2]. As they have no physical contact with the workpiece, friction and other forces are absent (Figure 1 b&c). Therefore, they can move at extremely high speeds limited only by their positioning systems. On the other hand, HED beams are used in focused conditions to have adequate energy density and hence have a very low diameter. Therefore, they need to make several passes to address any given area. Two broad applications of HED beams are:

i. Joining, cutting, engraving, etc.
ii. Additive Manufacturing (AM).

IOP Conf. Series: Materials Science and Engineering **759** (2020) 012012 doi:10.1088/1757-899X/759/1/012012

Table 2 summarizes the features of both these applications. While Gaussian energy distribution is favorable for the first group of applications, AM will require a uniform or flat hat shape as area-filling is essentially akin to paint-brushing (See Figure 2). While laser can achieve flat hat beam characteristic (i) optically or (ii) through oscillation, this is achievable only through oscillation in EB [1-4].

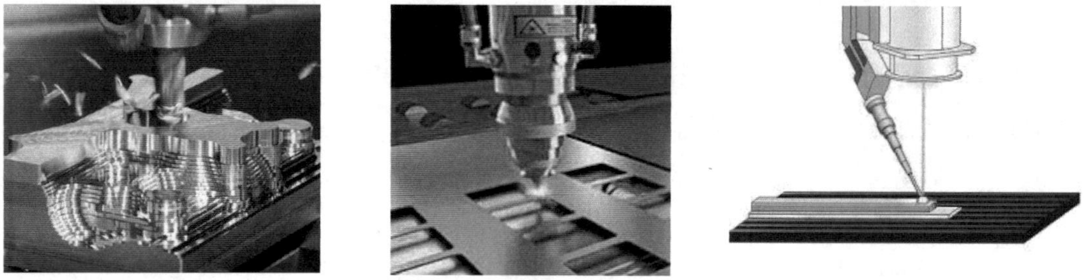

(a) (b) (c)

Figure 1. Contact and non-contact tools in manufacturing: (a) Contact tool in CNC machining; (b) Laser cutting; (c) Electron beam cladding. (b) and (c) are non-contact tools: HED beams

Table 1 Comparison of contact and non-contact tools in manufacturing

Feature	*Manufacturing Using*	
	Cutters	*HED Beams*
Physical contact and transfer of forces	Present	Absent
Speed of interaction	Extremely low	Extremely high
Diameter at interaction	Extremely high	Extremely low

Table 2 Comparison of the two broad applications of the HED beams in manufacturing

Feature	*Joining/ Cutting/ Engraving*	*Additive manufacturing (AM)*
Length of travel	Only contouring. So, short lengths.	Contouring as well as area filling. So, very long paths.
Number of passes of sintering/ Melting	Usually a single pass.	Multiple passes to gradual & uniform heating to minimize the residual stresses and warpages.
Motion accuracy	Low (0.080 to 0.100 mm will do)	
Speed of travel	Very less to allow sufficient time for melting – of the order of 1,000 to 5,000 mm/min (0.017-0.085 m/s).	Extremely high – 20 m/s possible for lasers using galvanometer; 10,000 m/s for EB using deflection coils.
Acceleration of travel	It can be very low – as a low as 5 m/s^2 will do.	It has to be extremely high due to short rasters.
Beam profile	Gaussian	Flat hat (uniform)

The build time in any 3D printing process using HED beam will be proportional to its volume. The part is built from thousands of slices. The regions in each slice defined by loops will have to be filled by moving the thin beam along millions of segments of sizes varying from a fraction of a millimeter to a few hundred millimeters. Furthermore, these motions may be straight lines (along X or Y or at any arbitrary orientation) or curvilinear.

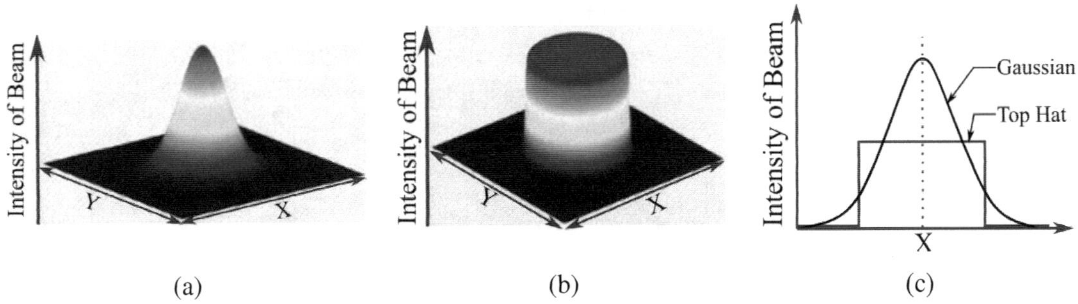

Figure 2. Gaussian and flat hat shapes of the HED Beams: (a) Gaussian; (b) Flat hat; (c) Both shown superimposed.

Table 3. Comparison of popular motion systems.

Motion system	Surface of focus	Spot size	Repeat-ability	Velocity	Acceleration	Application
Rotary motor with belt & pulley	Flat	Circular	±25μm	10 m/s	0.5g	Both. For lasers, the lens can be moved. For EB, the only table can be moved.
Rotary motor with leadscrew & nut	Flat	Circular	±15μm	2 m/s	0.5g	
Linear motors	Flat	Circular	±5μm	7 m/s	30-100g	
Galvanometers	Spherical	Oval	±15μm	10 m/s	> 5000g	Only for lasers
Deflection coils	Spherical	Oval	±15μm	> 10^4 m/s	> 10^4 g	Only for EB

Note: The values are the only representative for the purpose of qualitative comparison. 1 g = 10 m/s2 approx.

In summary, the thin HED beam in AM, capable of extremely high speeds and accelerations, has to be moved quickly along kilometers of distances constituted by short/long and straight/curvilinear segments. In order to achieve this, the following aspects deserve serious attention:

i. Motion system
ii. Path planning algorithm
iii. Scanning patterns and direction
iv. On/off switching of the beam
v. Beam shape
vi. Preference of hardware and analog systems to digital systems.

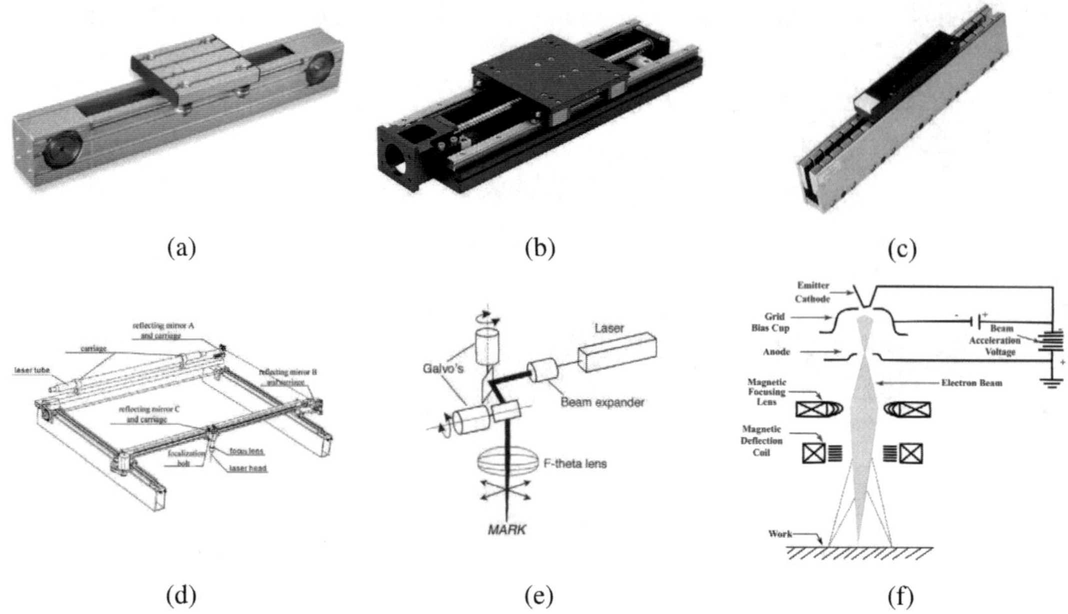

Figure 3. Popular motion systems: (a) Belt driven; (b) Screw driven: (c) Linear motor system; (d) Flying optics: (e) Galvanometer; (f) Deflection coils. (a) and (b) are rotary motor systems. (d) and (e) are motion controls specific to lasers. (f) is a motion control specific to EB.

2. Motion systems

Various possible motion systems are illustrated in Figure 3. Their performances are compared in Table 3. Consider a 20mm cube to be built by a beam of 0.5mm diameter in layers of 0.25mm. Assuming 75% stepover of a beam in consecutive passes, the total length of motion will be $\frac{20}{0.25}$layers x 20mm x $\frac{20}{0.5*0.75}$ passes = 85/333mm. Assuming a speed of 2m/s for a motion system using leadscrew, this can be made in about 43 seconds. But, this is under the assumption that the velocity is constantly at 2m/s. This is not possible in reality as the motor has to ramp up from rest to reach the maximum speed, cruise at that speed and then ramp down to rest again in each of the 20mm segments. Therefore, for an acceleration of 5m/s^2, the maximum speed will never be reached; it will be only $\sqrt{2 * 5 * 0.010}$ = 0.31623 m/s. So, the average speed will be half of this, i.e., 0.15811 m/s. So, although the maximum velocity is 2 m/s, because of the low acceleration, the average velocity is only 0.15811 m/s. Therefore, the building will take $\frac{85.333}{0.15811}$ = 550 seconds instead of just 43 seconds. In order to keep the focus on motion analysis, we choose to ignore here the inter-layer overhead time for table movement, material spreading, etc. which are substantial (10 seconds to 1 minute per layer). Since the motions will be often as small as 1mm, the acceleration of the motion system becomes the most important parameters in 3D printing. Since 3D printing has two inherent errors due to (a) stair-step way of growing and (b) support removal, repeatability of about 0.04mm is good enough. The maximum velocity of even 2m/s is adequate as long as it is achievable. As is seen from Table 3, the maximum velocities of all the motion systems listed are adequate. Therefore, the selection of the motion system in 3D printing using the HED beam is exclusively dictated by its acceleration.

Manufacturing requires relative motion which can be achieved either by moving the tool or the workpiece on the table. Table movement is usually through the rotary motor and hence will be slower. Therefore, it is desirable to move the HED beam wherever possible.

2.1. Motion system for lasers

Due to the low inertia of the tiny oscillating mirrors, the galvanometer is able to deflect the beam at extremely high speeds and accelerations. As illustrated in Figure 3 and Table 3, the galvanometer is the

IOP Conf. Series: Materials Science and Engineering **759** (2020) 012012 doi:10.1088/1757-899X/759/1/012012

best in terms of accelerations – over 5000g, which is about 100 times that of linear motors. But linear motors have accelerations higher by 50 times existing rotary motor systems. Therefore, they are a good compromise between galvanometers and rotary motors. We prefer linear motors to galvanometers because:

- Galvanometers are affordable only up to 1.5kW power which is well below future requirements in 3D printing of over 4kW power; efficient path planning algorithm is able to distribute uniformly the enormous heat without warpages.
- Galvanometers suffer spot geometry variations in size and shape (circle to ovality) at different (x, y) locations; correcting these errors optically is difficult and expensive.
- Galvanometers are more expensive than linear motors today.

2.2. *Motion System for EB*

As the deflection of EB is achieved by controlling the current flow in the coils, it is totally free from mechanical inertia [1]. Assume an EB of 1mm diameter at a height of 500mm from the current layer. It is possible to oscillate the beam over a range of ±30° at a frequency of 10 kHz. So, its scanning speed can be as high as $\frac{\pi}{3} * 0.5m * 2 * 10kHz \approx$ 10,000 m/s. Its acceleration too can be much greater than 10,000 m/s^2. These values are 1,000 times that of laser!

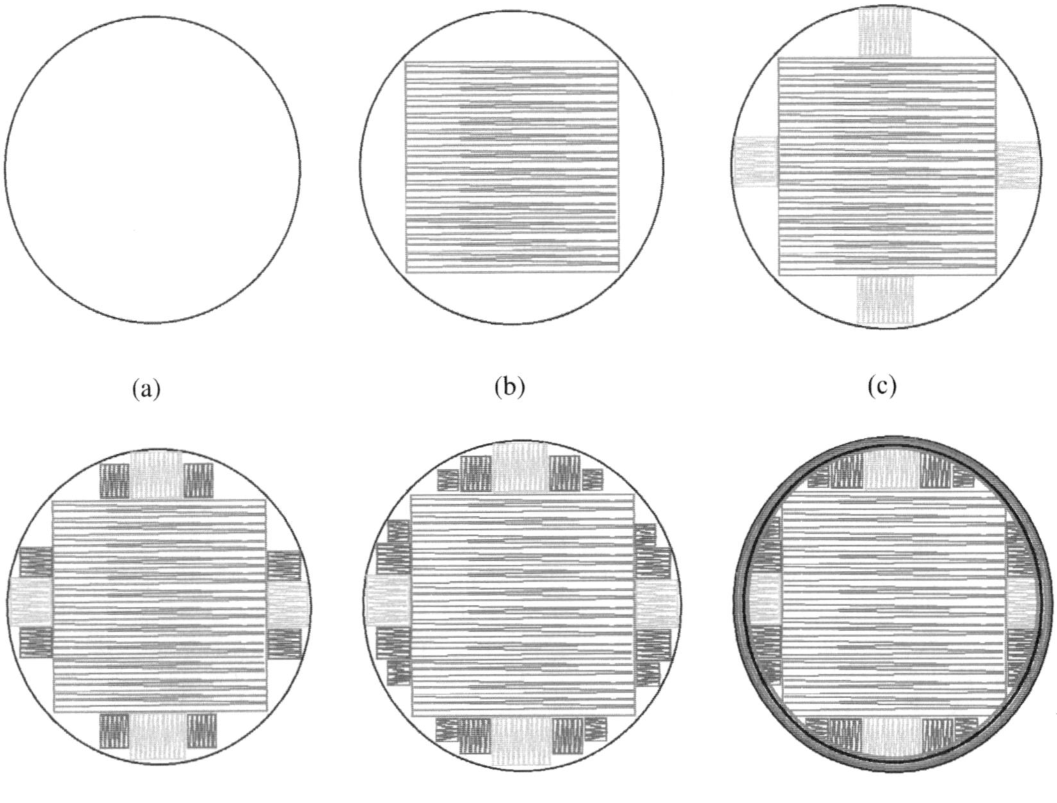

(a) (b) (c)

(d) (e) (f)

Figure 4. Area-filling using squares: (a) Circle to filled, (b) The 1st level of filling with squares; (c) The 2nd level of filling with squares; (d) The 3rd level of filling with squares; (e) The 4th level of filling with squares; (f) Final filling with contours.

While EB can be deflected at such phenomenal speeds, other related things too should match to exploit it. If we try to run EB along a boundary or fill it with raster lines, the digital computation of CNC and its interpolators comes into the picture. These digital computations take much higher time compared to the time scales that EB takes to cover a few millimeters. Therefore, we should replace digital computing with analog and hardware. So, the traditional area-filling strategies like *direction-parallel, contour-parallel, space-filling curves, traveling salesman algorithm,* etc. will not work for EB. Instead, the region should be divided into squares (or rectangles) of different sizes. Each square (x_c, y_c, s) shall be scanned using an analog circuit that generates appropriate saw tooth waves for X and Y coils and their biases to fill it. Finally, the contour itself may be offset by the required number of times to fill the rest of the area. Figure 4 demonstrates this area-filling concept for a circular region. Each square is filled in a sawtooth form is illustrated.

2.3. HED beam on/off

The on/off switching time of the HED beam is slow compared to the beam deflection speed. So, instant on/off does not happen to result in huge profile errors in area-fill. Therefore, we propose to switch on EB at the beginning of a layer and switch it off only after the entire layer is over. We implemented this for laser in sand 3D printing and found it satisfactory.

2.4. Scanning path and directions for HED beam

Any CNC interpolation takes time. Hence, even when the same speed is fed through a program, the results produced on the workpiece will be very different. This was very evident in the experiments carried out on thin sheet metal processing using lasers. As it is shown in Figure 5, while scanning over the diagonal of the rectangle, the speed of laser becomes less due to interpolation resulting in more heat input compared to sides of the rectangle. Even though appropriate offsets in speed during path generation can produce desired results, that leads to compromise over the utilization of speed we can achieve with HED beams. Furthermore, curvilinear motions of contour-parallel area-filling patterns suffer centrifugal forces in addition to higher computing time. Therefore, we use rasters along X and Y in alternate layers.

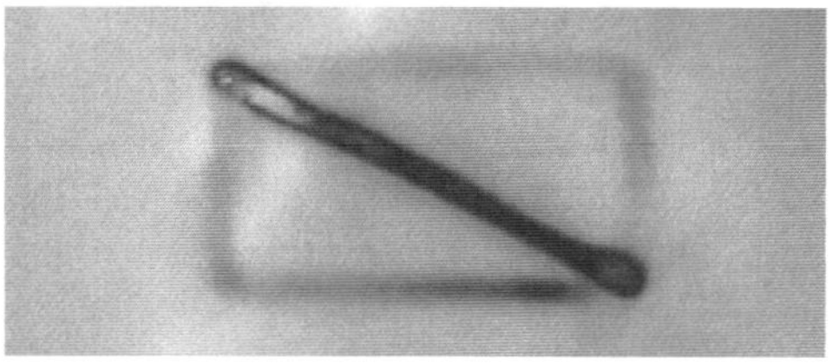

Figure 5. Effect of CNC interpolation on laser processing of thin sheets.

2.5. Segmentation of path of HED beam

The aspect ratio is the highest in 3D printing as compared to many manufacturing processes as building happens in thin layers. This results in warpages and residual stresses. Therefore, we have developed an algorithm that will segment the tool path into segments of certain fixed length, say 200mm; each segment is picked randomly and scanned. Thus, the heat input will be more uniform resulting in minimum residual stresses and warpages. Figure 6 demonstrates the segmented approach of area filling.

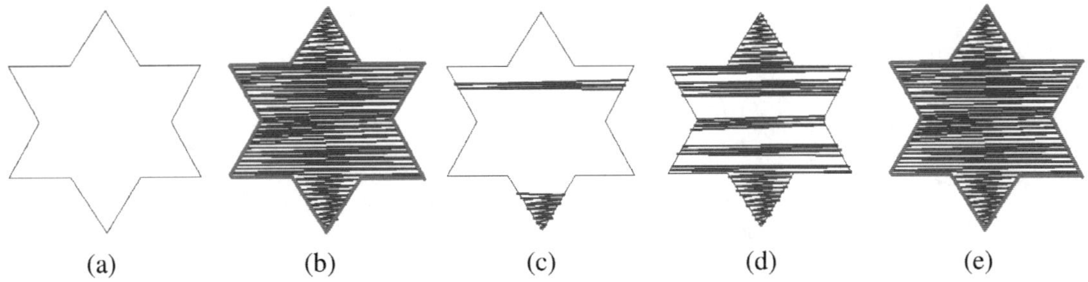

Figure 6. Random segmentation of area-filling motions to achieve uniform heat distribution: (a) Contour; (b) Area-filling path: (c), (d), (e) random writing of the segments.

3. Path planning to achieve flat hat beam profile and minimize build time

Although the concepts of this section are generic, the discussions here are in the context of laser beam positioning using flying optics driven by linear motors. The following are the two goals here:

i. Flat hat beam profile
ii. Minimization of scanning time

Flat hat profile is achieved in lasers optically using beam homogenizer. Another simpler way is to oscillate the beam in the transverse direction as shown in Figure 7a, which leads to consecutive Gaussian forms overlapping as in Figure 7b. As we move either on X or Y during area-filling, oscillation can be done along the other direction. The velocity of both the motors can be chosen such that about 75% stepover of the consecutive Gaussian forms will happen which is a reasonable approximation of flat hat.

However, this will mean that the motor in the oscillating direction (Y here) will be used more efficiently compared to that of the traveling direction (X here). We are keen to exploit both motors at the highest possible speeds so that the build time will be less and at the same time, the consecutive Gaussians will overlap to give a flat hat beam.

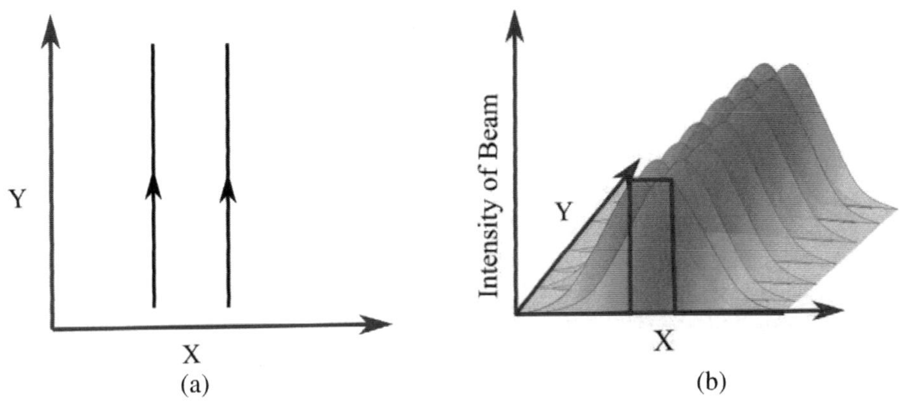

Figure 7. Oscillation to achieve flat hat: (a) Oscillation pattern; (b) overlapping Gaussian leading to flat hat

IOP Conf. Series: Materials Science and Engineering **759** (2020) 012012 doi:10.1088/1757-899X/759/1/012012

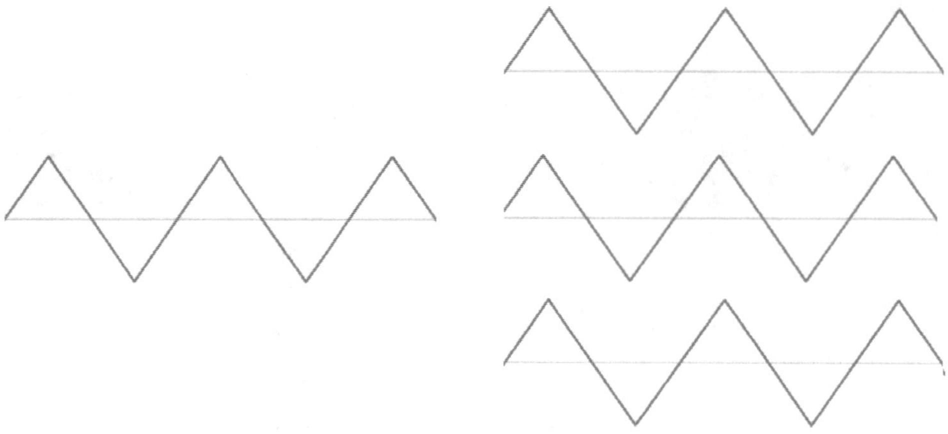

(a) (b)

Figure 8. Oscillation with equal speeds of the axes: (a) Single-pass; (b) Multiple passes with the flat hat.

When both axes are moved at maximum possible equal speeds, then, the result will be a triangular wave with a 45° angle as shown in Figure 8a. Instead of getting Gaussians overlapped between the consecutive oscillations as it is shown in Figure 7b, we shall tweak the starting and ending points of the area-fill raster so that the triangular waveforms of the consecutive passes match as shown in Figure 8b. This would achieve the required flat hat without any compromise on the build rate.

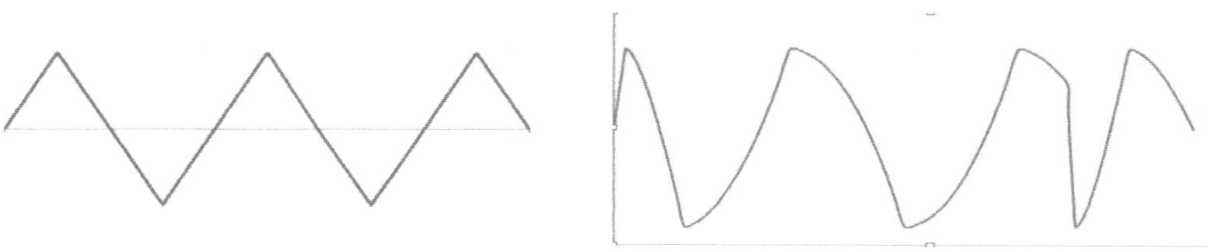

(a) (b)

Figure 9. Oscillation with equal speeds of the axes – (a) theoretical and (b) real paths.

The path shown in Figures 8a and 9a are theoretical assuming the same velocity for X and Y. However, their velocities may not be equal always. Furthermore, as the oscillation amplitude is less, the maximum attainable speed in Y-axis will be less than X-axis. Thus, the real path will be distorted as shown in Figure 9b.

This was tested in a case study presented in Figure 10 for a circle. Its rasters are shown in Figure 10a. When we used the exact points of the raster, the waves do not match as shown in Figure 10b and thus flat hat was lost. However, after appropriate tweaking of the endpoints, the waves matched as shown in Figure 8c. Note that the endpoints now fall in a grid of the working envelope. Therefore, area-filling is less complete. However, this can be handled by adding more contouring passes similar to the approach adopted in the square-based area-filling proposed for EB (Figure 4f).

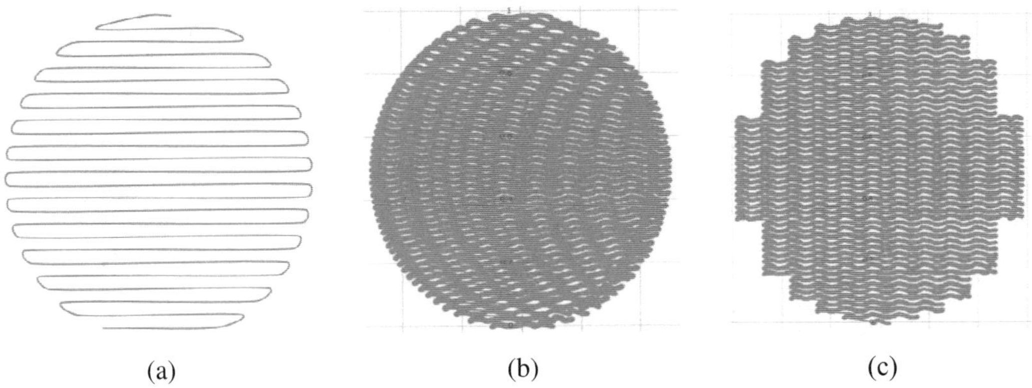

Figure 10. Case study of a circle: (a) Circle with rasters; (b) Area-fill without tweaking of the endpoints resulting in the loss of flat hat; (c) Area-fill with suitable tweaking of the endpoints resulting in a flat hat.

CAM packages are mostly geometric processors and rarely take into account the kinematics and dynamics of the process. In the 3D printing processes using any HED beam, we need to incorporate these parameters, viz., the maximum velocity and acceleration, during path planning.

4. Conclusions

The manufacturing speed in machining is limited by friction but it is limited by only the positioning system in 3D printers using the HED beam as there are no forces. As there are millions of tiny motions, the acceleration becomes the limiting factor than velocity or precision. The importance of a flat hat beam for 3D printing was demonstrated. In order to achieve a flat hat beam without any compromise on productivity, two path planning algorithms were proposed, one each specific to EB and laser, the former exploiting analog and hardware computing and the latter taking into account the maximum velocity and acceleration of the motion system.

References

[1] Schultz H 1993 *Electron beam welding* vol. 93 (Woodhead Publishing).
[2] Steen W M, and Mazumder J 2010 *Laser material processing* (Springer Science & Business Media).
[3] Gedicke J, Olowinsky A, Artal J, and Gillner A 2007. Influence of temporal and spatial laser power modulation on melt pool dynamics, In *Proc. of International Congress on Applications of Lasers & Electro-Optics* (October 2007, LIA), No. 1, p. 1502.
[4] Thiel C, Hess A, Weber R, and Graf T 2012 Stabilization of laser welding processes by means of beam oscillation *Laser Sources and Applications* **8433** 84330V.
[5] Luo X, Li J, and Lucas M 2017 Galvanometer scanning technology for laser additive manufacturing *Laser 3D Manufacturing IV* **10095** 1009512.

IOP Conf. Series: Materials Science and Engineering **759** (2020) 012013 doi:10.1088/1757-899X/759/1/012013

Control of electron beam wielding parameters based on the gap scanning system data during the welding process

A N Kasitsyn, R Yu Agafonov, A V Leonov, G V Grigoriev

Material science dept., Russian space systems, JSC, 53, Aviamotornaya ave., Moscow, 111250, Russia

E-mail: contact@spacecorp.ru

Abstract. Influence of the gap between the welded edges when sealing the case of electronic devices on the magnitude of the electron emission sensor signal in the process of electron beam welding (EBW) is studied. The impact of electron beam current variation at a constant value of the focusing system current on the penetration depth with the increasing gap between the welded edges is determined. The dependence of the length of the weld at different electron beam currents on samples with an increasing gap between the welded edges is studied. The possibility of controlling the process of electron beam welding by electronic emission from the weld pool is presented. A sensor for electronic emission registration is proposed, which includes an electron collector and an electron current data record system as a part of the EBW installation. The efficiency of the electronic emission sensor directly in the process of sealing under precision welding modes is shown.

1. Introduction

When sealing the cases of electronic devices using EBW in a vacuum chamber the required penetration depth should be provided. The design features of the device case may not provide for the minimum gap between the welded edges of the device case and its cover. Therefore, in the sealing process the non-fusion of the welded edges may occur due to a change in the welding gap size. At the same time, the elimination of such a defect leads to an increase in the penetration depth and, as a consequence, to a violation of the requirements for a welded joint.

In the process of the electronic devices cases sealing by EBW, it is possible to control the beam current, as well as its focus position and shape, using the information from various sensors to ensure the necessary penetration depth. Optoelectronic and X-ray sensors, as well as charged particle sensors based on electron collectors [1–3] are widely used to control the EBW process. The existing joint tracking systems are designed for the electron beam positioning in the direction of the welded parts joint and do not allow an assessment of the gap size between the welded edges in order to adjust EBW modes.

During EBW, due to beam electrons interaction with the welded edges, electrons kinetic energy transfers to thermal energy, causing metal heating and melting, as well as the appearance of various types of charged particles radiations. A certain part of electrons is reflected after the interacting with the metal surface. These electrons, which may have high or low energy, can give useful information about the processes occurring in the welding bath, and can be used in tracking systems and control systems for EBW. To register such information the electron collectors [4] are widely used. They are usually based on an electrode, which is grounded through current-measuring resistance and installed directly above the welding zone or close to it.

Using of electron collectors in EBW became widespread and is often used to find the joint of welded parts and adjust the position of the electron beam focus [5, 6]. Typically, a highly focused low-power electron beam is used in these systems to obtain information about the welded edges position by scanning them along different trajectories just before welding. In addition, several methods of EBW process control using plasma plume as a current conductor are applied [7], however, the sensors of such systems are located close enough to the weld pool and can be affected by significant thermal effects, which reduce the reliability of EBW process automatic control.

Before studying the influence of welding parameters on the penetration depth, the signal, which contains the information about the gap width, was singly researched. The signal was investigated using an electron emission collector under the influence of an electron beam on the welded edges. We also investigated the nature of the weld formation at different sizes of the joint gap.

2. Materials and methods

The system of electron current data record [8] consisting of a personal computer (PC) and an electron emission sensor based on an electron collector was used for the research. The schematic diagram of the electron current data record system as a part of the EBW installation is presented in figure 1.

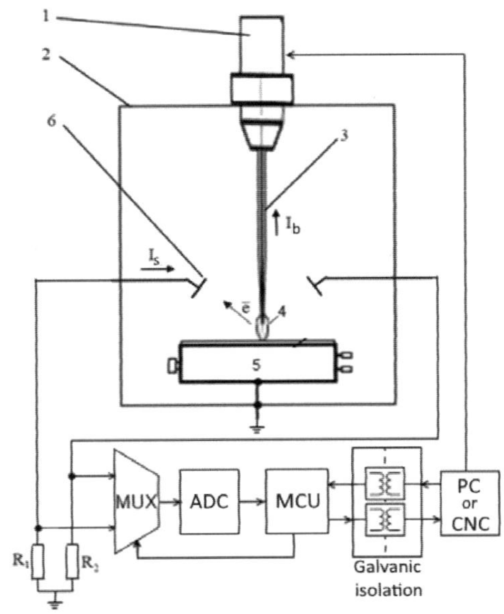

Figure 1. Schematic diagram of the electron current data record system as a part of the EBW installation: 1- electron beam gun; 2 - vacuum chamber, 3 - electron beam; 4 - plasma plume; 5 - case; 6 - electron collector; R1, R2 - resistance for matching the measured signal; MUX – multiplexer; ADC - analog-to-digital converter; MCU – microcontroller unit ; PC – personal computer; CNC - computer numerical control; I_b – beam current; I_s - signal current;

In the proposed scheme the electron collector 6 is an anode relating to the sealed case 5. A positive charge is supplied to the electron collector, so that the electrons emitted from the welded edges surface are attracted by collector 6. The generated electron current is determined by the sum of the thermionic current (thermal electron emission) and the current of reflected electrons (secondary emission), negative ions and positive ions.

Registration of electron current is performed by the electrode, which has the positive voltage of 5 V for its appropriate operation with the ADC. Electron current collected by the electrode passes through the 20 kohm resistance that connected to the ground (GND). The maximum current recorded by the proposed system is 350 μA at an electron beam current of 11 mA.

Thin samples with a length of L = 150 mm and a thickness of 1.3 mm were produced to perform an experiment to study the signal level and the electron current amplitude depending on the gap size. From one of the sides, the welded edges were tightly fixed in a snap without the gap. From the other side, the gap between the welded edges was provided by insertion of the 1 mm thickness plate. The gap sizes along the samples length was measured using a STEMI 2000-C stereo microscope. Welding was carried out at electron beam currents of 9, 10 and 11 mA. During welding, an electronic current was recorded using the electron emission sensor. The depth of penetration was estimated using the stereomicroscope on fragments, which were cut out from welded samples by machining and prepared for the research.

3. Research results and discussion

The graphs of electron current amplitude with the electron emission sensor during EBW of samples without gaps and with various gaps sizes were obtained. The analysis of these graphs allows noticing differences in signals from the electron collector, which could be useful for electron beam welding process control. In all cases, a moving average filter was used to reduce noise and highlight the signal [9]. During EBW without gaps the electronic current does not change significantly in time (see figure 2).

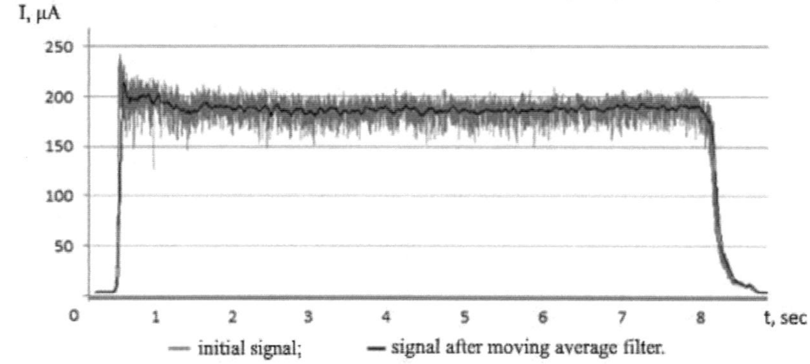

Figure 2. Electron current amplitude during the welding of the sample without the gap at a beam current of 9 mA.

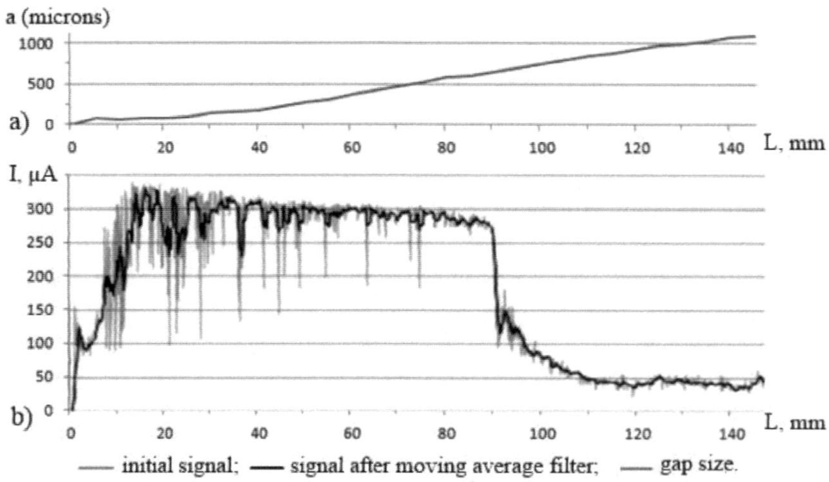

Figure 3. Electron current amplitude during the welding of the sample with a gradually increasing gap: a – the dependence of the gap width *a* from the of sample length *L*; b - a graph of the electron current amplitude *I* in the welding with the beam current of 11 mA.

During EBW with gap the electron current changes as the gap width increases that is shown in figure 3. The average value of the electron current decreases by 50 μA when the gap width increases to 750 μm. The amplitude of the electron current oscillations is also reduced. This reduction is caused by a gradual decrease in the number of electrons reflected from the edges surface and thermal electron emission from the weld pool. This process is possible because the part of the primary electrons of electron beam does not interact with the welded edges metal, but move into an increasing gap. The gap value becomes critical for the defined parameters of electron beam when a sharp decrease in the current amplitude occurs (approximately at a distance of L = 90 mm from the beginning). In that moment the fusion process stops and the melting of the welded edges begins separately. Further, as the gap increases, the current recorded by the electron collector continues to decrease. According to the graph "current amplitude – time" it is possible to determine after welding the coordinates of the non-fusion zone at a defined welding speed.

In case of the gap presence, a decrease in the signal amplitude can be observed in the middle of the welded samples, as can be seen in figure 4. It caused by an increase of the number of electrons falling into the gap region.

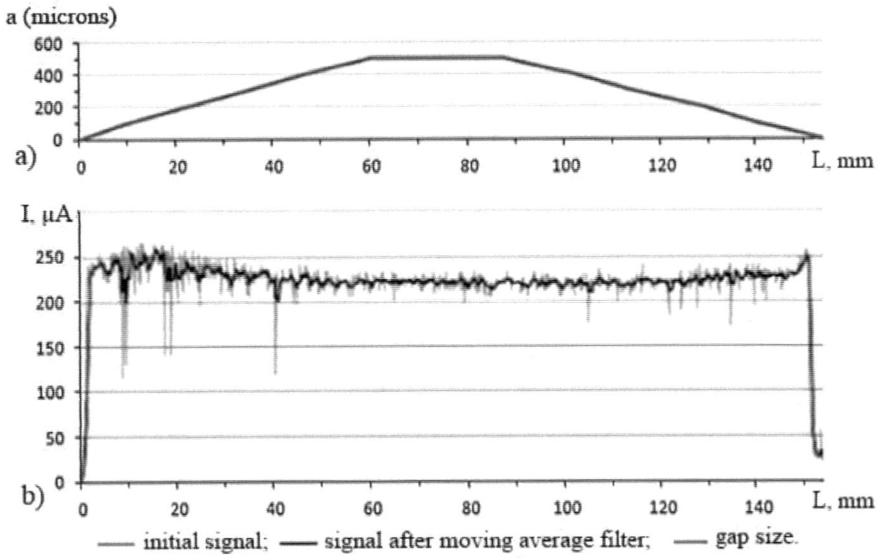

Figure 4. Electron current amplitude during the welding of the sample with a welding gap in the center: : a – the dependence of the gap width *a* from the of sample length *L*; b - a graph of the electron current amplitude *I* in the welding with the beam current of 11 mA.

In this case, non-fusion zone was not formed. However, over a long section of 60-120 mm, the average value of the observed electronic current is lower than at the beginning and at the end of the weld sample. This effect corresponds to an increase of welding gap up to 500 μm in the middle of the sample. As the gap decreases, the amplitude of the signal recorded by the sensor also increases.

It was experimentally established that the change in the beam current slightly affects the weldability of the edges with an increasing gap, which is reflected in figure 5. In all samples the non-fusion zones were occurred when the gap between welded edges was about 650-800 μm. Because of the current density distribution over the electron beam radius is Gaussian [10], the increase of power density in the heating spot due to the increase in current is not effective. The most efficient power distribution part of the beam falls into the welding gap and the energy put into the welded edges is not enough to maintain melting.

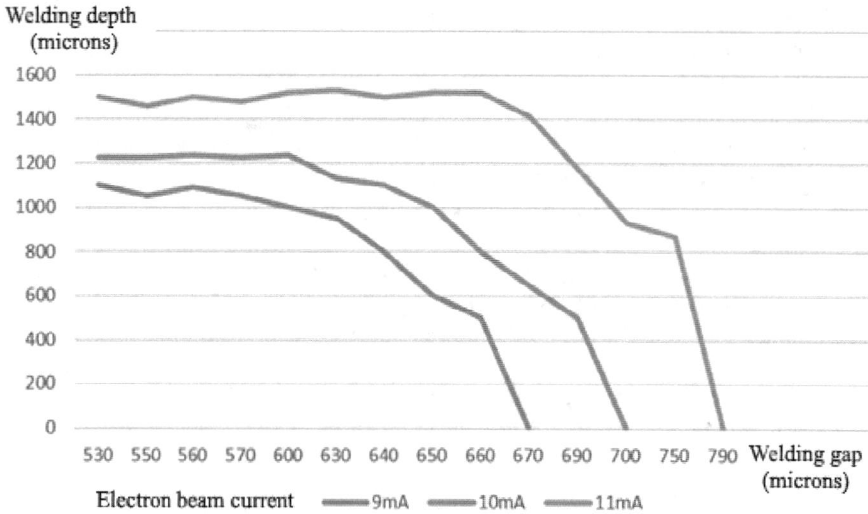

Figure 5. Dependence of the penetration depth on the gap width at different welding modes.

4. Conclusions

Based on the obtained relations, it can be concluded that the increase in electron beam current slightly affects obtaining of a welded joint with an increasing gap between the welded edges of the case and case cover of electronic devices. Despite the increase in the electron beam current, the power density in the heating spot drops significantly due to the hit of a part of the beam electrons in the gap.

In order to ensure the fusion of the welded edges under these conditions, it is necessary to change the current of the focusing system to increase the area of the heating zone. Together with the focusing current, the electron beam current should be increased to provide the power density required to form the weld pool. As a result of electron current graphs analysis, it is determined that the signal amplitude differs depending on the initial beam current, while the dependence character for determining the gap value is preserved.

To automate the process of controlling the penetration depth when sealing the elecronic devices cases, the EBW process control system based on information from the welding zone by using an electronic emission sensor during welding can be used. Comparing the reduction in electron current amplitude during the welding process, the automatic control system makes changes to the parameters of the EBW mode, allowing continuing the sealing process and obtaining a tight connection, satisfying the requirements. The scanning electron beam becomes relevant to use for welding of items with large gaps to ensure the fusion of welded edges.

5. Acknowledgments

The research was performed in the Material science department of "Russian space systems", JRC. Assistance with the literature searches and discussion platform was provided by National Research University "Moscow Power Engineering Institute".

References

[1] Sukach K A, Kutsan Yu G, Kovbasenko S N, Zhadkevich M L, Lankin Yu N and Zubov V V 1989 *A method for controlling electron beam welding and a device for its implementation.* 4257554/25-27

[2] Druzhinina A A, Laptenok V D, Murygin A V and Laptenok P V 2016 Automatic control of positioning along the joint during EBW in conditions of action of magnetic fields *IOP Conference Series: Materials Science and Engineering* **155** 012009

[3] Bashenko V V, Mauer K O and Mitkevich E A 1974 Using of reflected electrons to control the beam installation on the joint and the parameters of the joint in electron beam welding *Proc. LPI* **336** pp 97–100

[4] Druzhinina A A 2015 *Automatic compensation of the influence of magnetic fields on the positioning accuracy at the joint during electron beam welding* (Krasnoyarsk: Siberian federal university Press)

[5] Varushkin S V, Belen'kii V Ia and Trushnikov D N 2014 The use of shielded collector to control the mode of through-penetration in electron beam welding *J. Modern problems of science and education* **6**

[6] Piskunov A L, Lialin A N, Shchavlev V E and Abdullin A A 2012 Review of the results of the use of secondary emission signals to control and control the process of seam formation in electron beam welding *J. PNRPU Mechanical engineering, materials science Bulletin* **3** p 82

[7] Vinogradov V A, Shershnev N A, Pavlovskii A I and Grigor'ev Iu V 1993 *Method of regulating the process of electron beam welding* 1133781

[8] Kasitsyn A N, Agafonov R Yu and Grigor'ev G V 2017 Detection system of formation of defects in the process of electron beam welding of instrument housings of radio-electronic equipment of spacecraft *Conf. Electron beam welding and related technologies* (Moscow: National Research University "Moscow Power Engineering Institute") pp 458-467

[9] Kir'ianov, D V and Kir'ianova E N 2006 *Computational physics* (Moscow: Polybook Multimedia)

[10] Rykalin N N, Uglov A A, Zuev I V and Kokora A N 1985 *Laser and electron beam processing of materials* (Moscow, Engineering Publ) p 496

Study on formation of aluminum alloy thin wall produced with WAAM method under various thermal conditions

A V Kisarev[1], N V Kobernik[2]

[1] S7 R&D Center, 5, Vostochnaya st., Gorki Leninskiye, 142712, Russia
[2] Bauman's Moscow State Technical University, 5a, 2nd Baumanskaya st., Moscow, 105005, Russia

E-mail: a.kisarev@s7.ru , koberniknv@yandex.ru

Abstract. The study reveals some formation aspects of a thin walls made by WAAM technology. The series of experiments about thin wall deposition in a various thermal condition was carried out. Temperature field finite element model was developed and solved under the giving conditions. Experimental results were treated after metallographic analysis, functional dependences of the width and height of the deposited layer by the temperature of the previous layer were constructed. In a range of 30-300 C sublayer temperature, layer height and width values show linear functional dependencies by sublayer temperature.

1. Introduction

WAAM (Wire Arc Additive Manufacturing) is a method of producing parts layer-by-layer by metal deposition using filler material in a form of welding wire. Both electric and compressed electric arc can be used as a heat source for this process. One of the key features in this technology is an ability to use standard welding equipment, such as MIG\MAG welding power source. In most of the cases WAAM is used to produce large parts from aluminum or titanium alloys.

WAAM has several advantages over the other metal 3D printing technologies. In comparison with selective laser sintering (SLM) method, this technology has much higher productivity rate and can significantly expands size of the parts to be printed[1]. Compared to direct energy deposition (DED), WAAM allow using welding wire instead of metal powder as a deposit material, which led to easier material quality control and also benefited in material availability and cost[2] [3].

Nowadays a lot of studies are devoted to WAAM implementation in various industries like shipbuilding or aerospace[4] [5] [6]. The main focus most of researcher teams in this process are pulsed arc technologies, like cold metal transfer (CMT) method[7] [8] [9]. CMT is a pulsed short-circuit welding technology developed by Fronius Inc. (Austria), which ensures deposition with controlled droplet transfer by a special welding parameters pattern using feedbacks, the droplet transfer occurs at 50-60 Hz frequency[10]. In addition, CMT provides a significantly lower heat input in base metal compare to conventional MIG\MAG welding methods and also has minimal spatter during the deposition[11] [12]. Furthermore, CMT can significantly increase stability of WAAM process and ensure the deposited material mechanical properties to be close to annealed state[13] [14].

However, during the deposition with constant process parameters (such as welding current, arc voltage, travel speed and wire feed speed) various layers formation can be observed, particularly in width and height values (fig. 1). The main cause of this phenomena is the thermal saturation of already completed previous layers. The new layer is depositing on previous preheated layer, which changes the spreading and formation condition of the material. This condition state can be described by the temperature of previous layer, i.e. sublayer temperature.

Finding influence degree of sublayer temperature on layers geometry is a very important step to ensure steady printed wall formation, especially for so-called thin walls. Thin walls (fig. 2 c) are such constructs that are made without welding torch oscillation (fig. 2 a) or multilayer deposition (fig. 2 b).

Therefore, the aim of this study is to determine the influence of temperature conditions on formation of a thin aluminum alloy wall and its shape parameters.

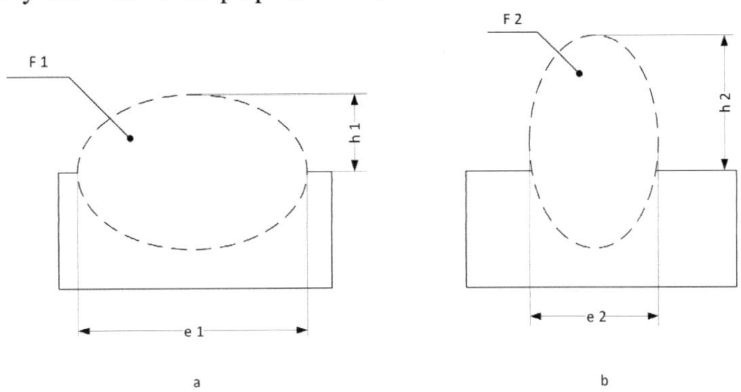

Figure 1. Layer formation under different thermal conditions: a - hot sublayer; b - cold sublayer

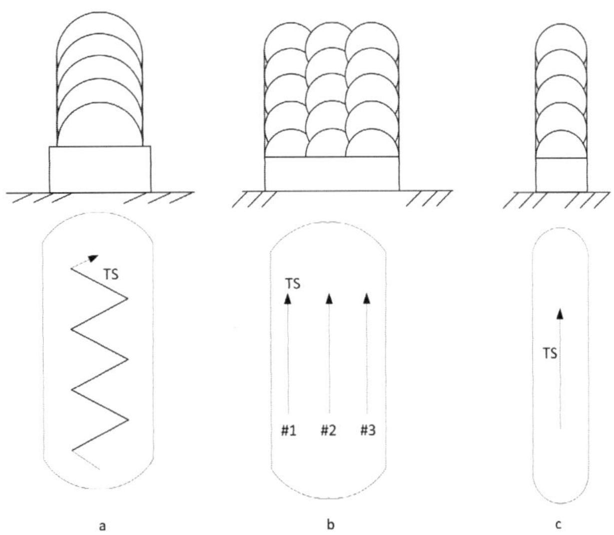

Figure 2. Wall deposition patterns: a – wall with weaving; b – multilayer wall; c – thin wall; TS – travel speed of the heat source

2. Materials and methods

In this study, the formation of a thin wall made by WAAM method using CMT technology was investigated depending on the temperature of the sublayer. Fronius TPSi 500 CMT welding machine was used as an energy source, Motoman MA2010 industrial robot represented the welding torch moving mechanism, ER5556A welding wire and argon shielding gas were user as consumables. The deposition was occurred from substrate plate of ER5356 aluminum alloy, 6x300x140 mm in dimensions.

IOP Conf. Series: Materials Science and Engineering **759** (2020) 012014 doi:10.1088/1757-899X/759/1/012014

During the experiment, the robot moved with constant speed along a linear path, rising through each layer for equal distance to maintain a constant electrode extension length. After each wall was finished, the reform was cooled to room temperature before proceeding to the next experiment.

Since the process of deposition is characterized by formation defects the starting and finishing areas, direction of the deposition pattern was alternating, as it is shown in Figure 3. Thus, these formation defects were compensated by each other during the deposition of the next layers.

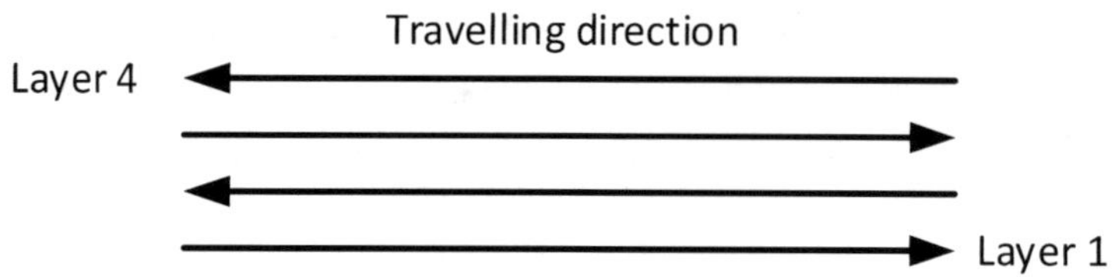

Figure 3. Deposition pattern

To simulate different sublayer temperature effect during the experiment, the time between deposition passes (layers) was varied, thereby changing the sublayer temperature. Thus, with different times between layers, the previous layer cools to different temperature from one experimental sample to another, but within particular sample each layer cools the same time. In order to reduce the substrate influence on the heat transfer, the first two layers in each sample were deposite with the same process parameters regardless of the sample number and were not considered in further analysis.7 samples in total with 30 layers each were performed on different welding parameters, presented in table 1 and table 2.

Table 1. Deposition parameters

Parameter	Value
Wall length	140 mm
Total layers amount	30
Travel speed, layer #1	0,24 m/min
Heat input, layer #1	1,625 kW
Travel speed, layer #2	0,36 m/min
Heat input, layer #2	1,1 kW
Travel speed, all other layers	0,6 m/min
Heat input, all other layers	0,67 kW

Table 2. Pause time between layers

Sample #	Pause time between layers #3-30, sec
1	2
2	7
3	12
4	30
5	60
6	120
7	240

After the experiment, the samples were cut out according to the scheme shown in Figure 4. Total of 28 cutouts were obtained to collect geometric parameters of the layers. Images were obtained from each cutout using optical microscopy, then were analyzed in the ImageJ image processing software. Height was calculated for each image by division of total height value (except first two) by the number of layers.

Based on the results of measuring shapes geometry, a finite element model was made in ANSYS software, simulating the temperature field during walls processing. In the calculation, the sublayer temperature in a point located directly below the heat source at a time corresponding to its passage through the center of the sample was taken into account. Normally distributed Gaussian source was used as a heat source model. Verification of the model was carried out using thermocouples with NI 9212 module (manufactured by National Instruments, USA, measurement accuracy of 0.71 ° C) as a controller. Thermophysical properties of the material were set from the material library of ANSYS simulation software.

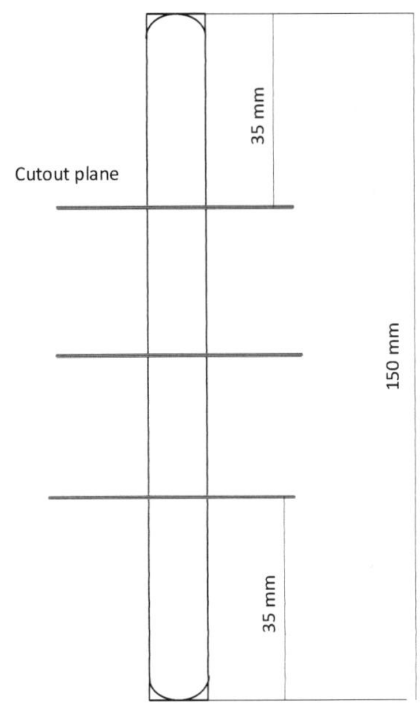

Figure 4. Cutout scheme for geometrical parameters analysis

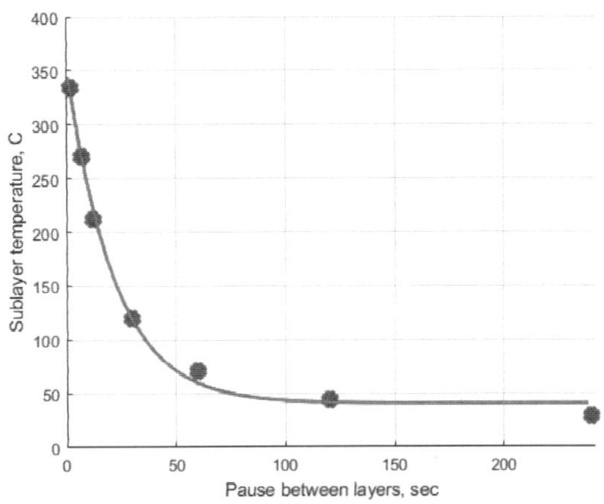

Figure 5. Results of numeric modelling: sublayer temperature by pause between layers

According to the results of a numerical experiment, the dependence of the sublayer temperature for a point in the center of the sample by the time between the layers was obtained as shown in Figure 5.

3. Results and discussion

The results of layers geometric parameters measurements are shown in Table 3 and in Figures 6 and 7.

Table 3. Measurement results.

Sample #	Pause between layers, sec	Sublayer temperature, °C	Mean height of a single layer, mm	Height measurements standard deviation, %	Mean width of a single layer, mm	Widht measurements standard deviation, %
1	2	334	1.51	1.02	5.98	2.85
2	7	270	1.60	1.42	5.46	0.55
3	12	212	1.65	0.98	5.37	1.79
4	30	120	1.78	0.29	4.36	12.43
5	60	71	1.83	0.45	4.48	2.38
6	120	44	1.86	0.34	4.30	0.81
7	240	29	1.89	0.53	4.12	6.781

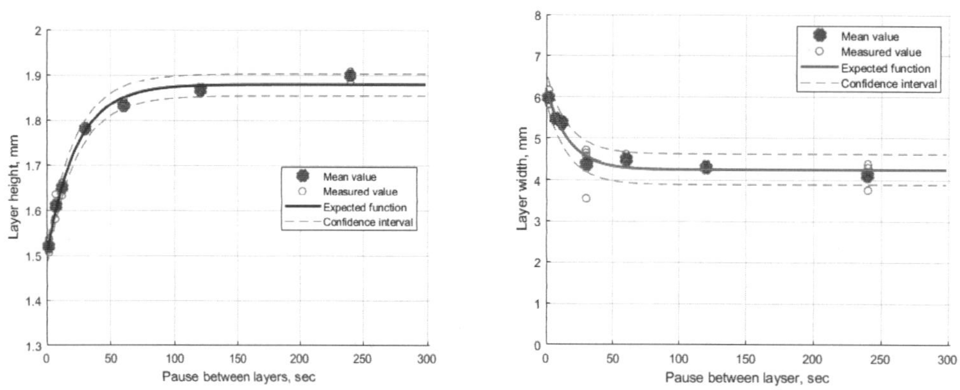

Figure 6. Measurements results: layer height (a) and width (b) dependency by pause between layers

Based on the obtained data processing, linear dependencies in a form of $f(x) = ax + b$ were constructed using the least squares method. The coefficients are given in table 4.

Table 4. Linear functions coefficients.

	a	b
Layer height	-0.0012	1.9229
Layer Width	0.0059	-3.9572

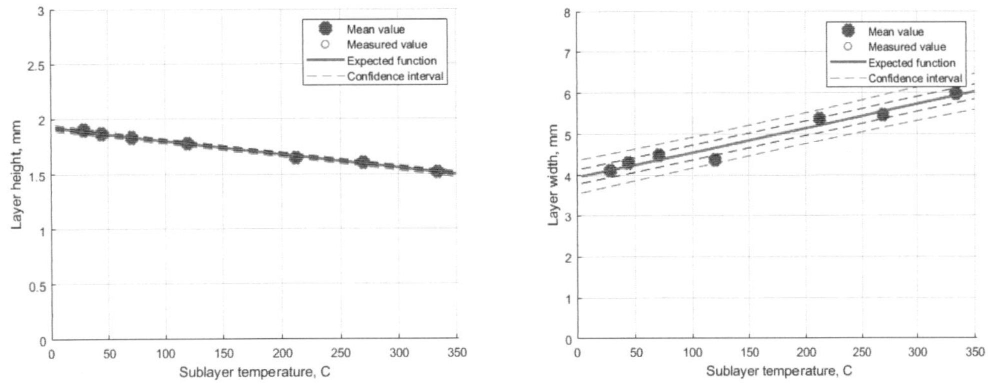

Figure 7. Experiment results: layer height (a) and width (b) dependency by sublayer temperature

Thus, the dependences of the height and width of each layer are established depending on the temperature of the sublayer, which must be taken into account when developing technology for printing any large parts. Predictive modeling of the process can help forecast the real geometry of the part, but it will take considerable time and require significant calculating resources. Contactless tracking methods seem to be the most convenient way of monitoring natural sublayer temperature. Their implementation will allow real-time data analysis and, if there is a feedback process control system, real-time adjustment of the process parameters in such a way as to obtain the desired shape of the part. However, it is necessary to verify the obtained dependencies when changing the general parameters of the process during real-time control.

4. Conclusions

1. The sublayer temperature has a significant effect on the geometry of each layer during WAAM process.
2. It can be assumed that in the temperature range from 30 to 300 ° C for certain deposition parameters, the dependence of the width and height by the temperature of the sublayer is linear.
3. The use of a real-time temperature monitoring system and organize feedbacks for main parameters of the deposition process might ensure stability of the thin wall geometry.

References

[1] Xu X, Ding J, Ganguly S, Diao C and Williams S 2018 Oxide accumulation effects on wire + arc layer-by-layer additive manufacture process *J. Mater. Process. Technol.* **252** 739–50

[2] Horgar A, Fostervoll H, Nyhus B, Ren X, Eriksson M and Akselsen O M 2018 Additive manufacturing using WAAM with AA5183 wire *J. Mater. Process. Technol.* **259** 68–74

[3] Prado-Cerqueira J L, Camacho A M, Diéguez J L, Rodríguez-Prieto Á, Aragón A M, Lorenzo-Martín C and Yanguas-Gil Á 2018 Analysis of favorable process conditions for the manufacturing of thin-wall pieces of mild steel obtained by wire and arc additive manufacturing (WAAM) *Materials (Basel).* **11**

[4] Jaffiol C and Baldet L 1984 Wire arc additive manufacturing of Al-6Mg alloy using variable polarity cold metal transfer arc as power source *Ann. Med. Interne (Paris).* **135** 359–64

[5] Derekar K S 2018 A review of wire arc additive manufacturing and advances in wire arc additive manufacturing of aluminium *Mater. Sci. Technol. (United Kingdom)* **34** 895–916

[6] Pang J, Hu S, Shen J, Wang P and Liang Y 2016 Arc characteristics and metal transfer behavior of CMT + P welding process *J. Mater. Process. Technol.* **238** 212–7

[7] Krampi M N Modern methods for pulse wielding developed abroad *Int. J. Appl. Fundam. Res.* 18–21

[8] Ahsan R U, Kim T, Kim D, Ji C and Park Y 2018 A Study on the Effect of Wire Composition on Welding with Gap and Offset in Cold Metal Transfer (CMT) GMAW 1–7

[9] Pickin C G and Young K 2006 Evaluation of cold metal transfer (CMT) process for welding aluminium alloy *Sci. Technol. Weld. Join.* **11** 583–5

[10] Cong B, Ouyang R, Qi B and Ding J 2016 Influence of Cold Metal Transfer Process and Its Heat Input on Weld Bead Geometry and Porosity of Aluminum-Copper Alloy Welds *Rare Met. Mater. Eng.* **45** 606–11

[11] Wang P, Hu S, Shen J and Liang Y 2017 Characterization the contribution and limitation of the characteristic processing parameters in cold metal transfer deposition of an Al alloy *J. Mater. Process. Technol.* **245** 122–33

[12] Dutra J C, Gonçalves e Silva R H and Marques C 2015 Melting and welding power characteristics of MIG–CMT versus conventional MIG for aluminium 5183 *Weld. Int.* **29** 181–6

[13] Imoudu N E, Ayele Y Z and Barabadi A 2018 The characteristic of cold metal transfer (CMT) and its application for cladding *IEEE Int. Conf. Ind. Eng. Eng. Manag.* **2017-Decem** 1883–7

[14] Ma G, Zhao G, Li Z, Yang M and Xiao W 2019 Optimization strategies for robotic additive and subtractive manufacturing of large and high thin-walled aluminum structures *Int. J. Adv. Manuf. Technol.* **101** 1275–92

IOP Conf. Series: Materials Science and Engineering **759** (2020) 012015 doi:10.1088/1757-899X/759/1/012015

Analysis of the amplitude-time parameters of current pulses in a plasma during laser beam welding

I Yu Letyagin[1], V Ya Belenkiy[1], D N Trushnikov[1] and S Pang[2]

[1] Perm National Research Polytechnic University, Perm, Russian Federation
[2] Huazhong University of science and technology, Wuhan, China

E-mail: letyagin@pstu.ru

Abstract. The share of application of laser processing of materials in technological processes of mechanical engineering has increased over the past few years. In this regard, there is a need for reliable control of the technological parameters of the laser welding process. In deep penetration laser welding, part of the beam power is absorbed by the plasma cloud. As a result, a significant decrease in the penetration depth is observed. Using amplitude-time pulse of current in the plasma in the zone of impact of the laser beam on the metal is one of the directions of the work necessary for the control of the processes in the melting channel. The study of the processes of secondary emission in the plasma in the zone of impact of the laser beam on the metal allowed carrying out numerical simulation of the processes during laser welding. The models took into account the dependence on the focusing of the laser beam and other technological parameters of laser welding. The registration of the secondary emission current was also carried out in order to control the geometric parameters of penetration during laser welding. This technique can be used to build methods for the operational control of the welding process.

1. Introduction

The application of laser material processing continuously expands in high-technology industries to produce high quality welds of parts of the structural steels and non-ferrous metal alloys.

As a first source of high-power laser radiation, a carbon dioxide laser was used for welding. They had large dimensions, low (not exceeding 20%) efficiency and could not compete with electron-beam welding equipment.

The emergence of high-power fiber-optic process lasers has contributed to the wider introduction of laser welding into industrial production. These lasers can be compared to carbon dioxide lasers. Lasers have higher reliability, relatively small size and provide high optical quality of radiation. In addition, the radiation wavelength of fiber-optic lasers is an order of magnitude smaller than that of carbon dioxide lasers. This causes less energy consumption. Their continuous improvement ensures lower cost and increased efficiency of technological laser systems. Powerful fiber-optic lasers have drawbacks: the operational reliability of the resonator decreases with increasing laser power, the sensitivity of the resonator increases to beam reflection processes, which arise as a result of the interaction of laser radiation with the material being processed.

In recent years, as high-power industrial lasers, disk lasers began to be applied. The principle of operation of these lasers is based on the use of cooling the active element in the form of a disk. High cooling efficiency of the laser medium is provided due to the large surface area of the disk. Consequently, the radiation power in a disk laser beam can reach quite high values. Important advantages of disk lasers are the ability to control the radiation power without changing other parameters, the lack of sensitivity of the resonator to the reflected laser radiation, as well as the modular design of the laser, which allows replacing individual modules during service.

Currently, one of the major problems in laser welding with deep penetration is the absorption of a part of the beam power by a plasma cloud formed above the zone of laser beam action on the metal. This is associated with a significant decrease in the depth of penetration compared to the electron beam of the same power. The efficiency of laser welding when joining thick-walled parts is significantly inferior to the process of electron-beam welding, which is widely used in the manufacture of critical products at the same time. This hinders the widespread introduction of laser welding in industrial production.

Certain technological methods are used to reduce the screening effect of the plasma cloud and accordingly to increase the penetration depth during laser welding. Technological methods are known: the application of special coatings on the metal before welding; the deviation of the axis of the laser beam from the normal to the side opposite to the welding direction by 20°...30°; oscillations of the laser beam in different directions relative to the joint; the use of a pulse-periodic mode of the laser; blowing the gas-plasma cloud with various gases and their mixtures [1-5].

However, these technological methods do not solve the problems associated with the power loss of the laser beam in the plasma cloud over the welding zone. Interest in laser welding in vacuum has increased in recent years. This method of welding has some limitations associated with the use of a vacuum chamber. It allows getting a much greater penetration depth at the same power of the laser beam as compared to gas-shielded laser welding. The use of a vacuum chamber provides effective protection of the welding zone from the external environment, which is especially important when welding active metals [6-8].

Operational control of the weld formation process can ensure high reproducibility of the quality of welded joints and the absence of defects in the weld seam. Secondary emission methods for controlling the process of the interaction of an electron beam with a metal have found wide application in electron beam welding. The parameters of the secondary emission signals from the zone of interaction of the electron beam with the metal are recorded with these methods of control [9-12].

Processes similar to processes in electron beam welding with a high-power electron beam occur in the area of laser welding in a vacuum. The following processes are known: intense thermionic emission from the condensed metal phase, the formation of a plasma cloud over the welding zone and the presence of a wide range of oscillatory processes in the melting channel formed by a powerful concentrated laser beam [13-15].

The purpose of this work is to analyze the amplitude-time parameters of plasma current pulses in the zone of influence of a powerful laser beam on a metal surface. Experimental work is carried out to evaluate secondary emission processes in the laser welding zone in order to use these signals to monitor and control the process of weld formation.

2. Research and experiments

The registration of the secondary emission current in the plasma in the laser welding zone was carried out by installing a collector of charged particles above the welding zone (Figure 1). It was included in an electrical circuit containing a load resistor and a source of bias voltage. The signal from the load resistor was fed to a computer information-measuring system with an analogue-digital interface based on an analogue-digital converter U-14-140M of the company L-CARD.

The experiments were carried out on the laser welding machine model ALFA-300, supplemented by a vacuum chamber with variation of the maximum drive voltage from 200 V to 400 V, pulse duration from 4 ms to 20 ms and their repetition rate 1 Hz.

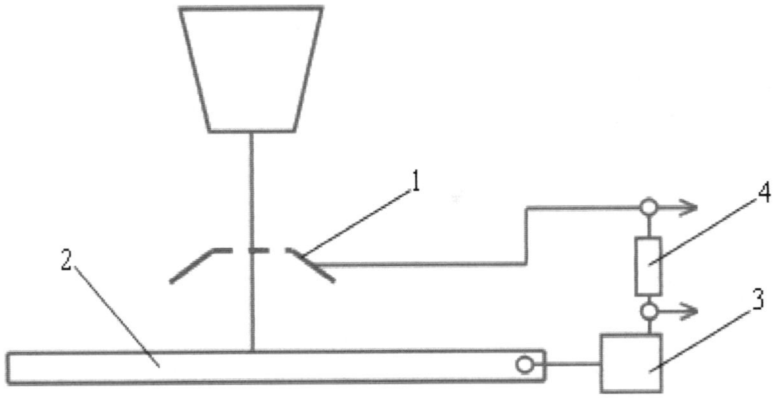

Figure 1. Registration scheme of the secondary emission current during laser welding in vacuum: 1 – collector of charged particles; 2 – welded product; 3 – source of bias voltage; 4 – load resistor.

Experimental studies have shown that the spectrum of the recorded signal contains a high-frequency component (f > 10 kHz) and its harmonics (Figure 2).

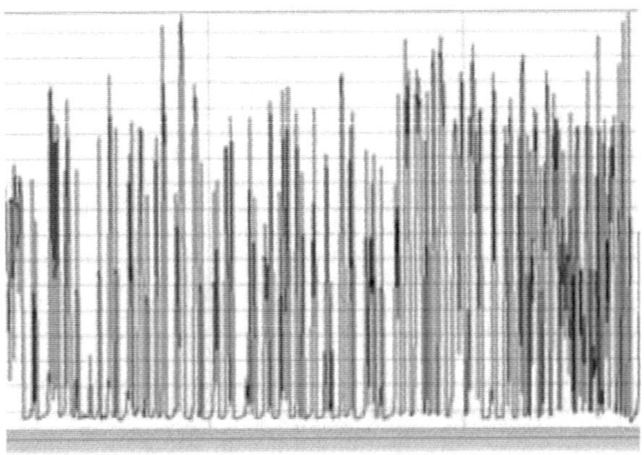

Figure 2. Fragment of the oscillogram of the secondary emission current in the plasma recorded by the collector of charged particles during laser welding.

This component of the secondary current in the plasma was measured using a collector of charged particles; it reflects the oscillatory processes in the plasma column when applied to the collector of positive potential.

The nature of these self-oscillation processes is associated with the onset of ion-acoustic instability. The observed instability is similar to the potential-relaxation instability, characterized by large amplitude of oscillations on the positive electrode located in the plasma. Both types of instabilities have similar excitation and propagation nature. These instabilities in the plasma appear when the discharge current density exceeds a certain critical value. The critical current density j, is expressed through the threshold value of the drift velocity of electrons

$$V_D = \frac{j}{n_e e}, \tag{1}$$

where n_e – the electron concentration, e – the electron charge.

The minimum value of the velocity V_D at which ion-sound instabilities arise coincides with the velocity of ion sound in plasma

$$V_s = \sqrt{(\gamma_e k_b T_e + \gamma_i k_b T_i)/m} \tag{2}$$

where k_b – the Boltzmann constant; γ_e – the coefficient, which for electrons is taken to be unity, for ions, $\gamma_i = 3$; T_e and T_i – respectively, the temperature of electrons and ions; m – the mass of ions.

The critical current density is ~ 3 mA / m^2 , with an ion temperature above the welding zone of about 2400 K, an electron temperature of ~ 10000 K and a plasma density $n_e \approx 10^{16}$ m^{-3}, which corresponds to a collector current of charged particles of about 4 mA for an area of about 2...4 cm^2.

The experimentally observed pulses of the electron current in the plasma have a magnitude greater than the threshold value.

Dispersion ratio for ion-sound waves is

$$\omega(k) = kV_s / \sqrt{1 + k^2 \lambda_D^2} \tag{3}$$

where ω – the frequency, k – the wave number, V_s – the phase velocity of the ion-sound waves in the plasma, λ_D – the Debye screening radius.

The dispersion relation takes a linear form $\omega(k) = kV_s$ for ion-sound oscillations in the case of large wavelengths ($\lambda >> \lambda_D$). The speed in ion-acoustic waves is constant, and the frequencies can take on a wide range of values, depending on the wavelength. This is possible if the ionic oscillations of the Langmuir are characterized by a constant frequency and arbitrary velocity values. If we take the quantity L/n as the wavelength of the n-th harmonic (L is the characteristic size of the system, i.e. the distance between the metal surface and the collector of charged particles), we obtain the expression for harmonic frequencies

$$f_n = nV_s / L \tag{4}$$

The frequency of the high-frequency component depends on the distance between the weld zone and the collector of charged particles. This agrees well with the experimental data. In this case, the evaluation of the phase velocity of the corresponding ion-acoustic waves gives a value of 1650 m/s.

Arising self-oscillations in the plasma above the welding zone modulate the oscillations in the spectrum of the secondary signal in the range of 100 Hz – 10 kHz. They arise due to: capillary instability of the penetration channel, stochastic movement of the zone of interaction of the laser beam with the metal on the walls of the penetration channel, local overheating in the penetration channel, pulsations of steam flow from the penetration channel and other periodic processes in the penetration channel.

The mechanism of collisional damping of oscillations is known in plasma physics. The probability of collisions of plasma particles during the oscillation period increases in the zone of laser welding at atmospheric pressure at high concentrations of neutral atoms. The number of these collisions is large enough, and the oscillations decay very quickly.

The task was to differentiate the oscillatory processes in the plasma associated with various types of instabilities and self-oscillations caused by periodic processes in the channel of melting. It is assumed that self-oscillations should correlate to a greater degree with the technological parameters of laser welding in vacuum and with the geometry of the melting zone in the metal when exposed to a laser beam.

The studies were conducted with the registration of the signal of amplitude-time parameters of the current of the secondary emission. The influence of the distance between the collector of charged particles, which produces a secondary current from the plasma and the surface of the metal exposed to the laser beam, was evaluated. This made it possible to take into account the nature of self-oscillations in plasma associated with various instabilities largely determined by the geometry of the plasma cloud formed in the zone of laser welding in vacuum.

Figure 3 shows the dependence of the frequency of the spectrum component due to ion-acoustic oscillations in the plasma in the zone of laser welding in vacuum, on the distance between the charged particles collector and the metal surface.

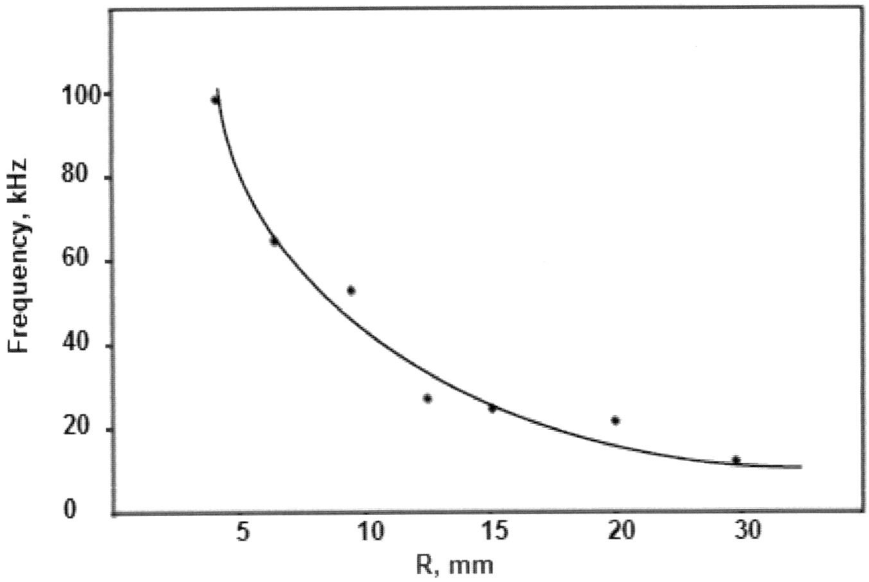

Figure 3. Dependence of the frequency of the spectrum component due to ion-acoustic oscillations in the plasma in the zone of laser welding in vacuum on the distance between the charged-particle collector and the metal surface

Figure 4 shows the oscillation spectra obtained by mathematical processing in the MathCad medium of the results of recording the signals of secondary emission in the zone of laser welding in vacuum. The data correspond to different distances from the collector of charged particles to the metal surface.

The spectral density of the high-frequency components of the secondary-emission current spectrum in the laser-welded zone decreases in vacuum, and their frequency increases as the distance from the metal surface to the collector of charged particles decreases, as can be seen from Figure 3 and Figure 4. Consequently, the high-frequency components of the spectrum of the secondary emission current in the zone of laser welding in vacuum are associated with ion-acoustic and relaxation instabilities in the plasma.

This proves that the electrode-collector of charged particles must be placed in close proximity to the surface of the welded metal. This will ensure the registration of the secondary emission current in order to control the geometric parameters of penetration during laser welding. Additionally, in the course of the experiments, an experimental dependence was obtained, which determines the extreme nature of the change in the magnitude of the secondary emission signal depending on the focusing of the laser beam and other technological parameters of laser welding.

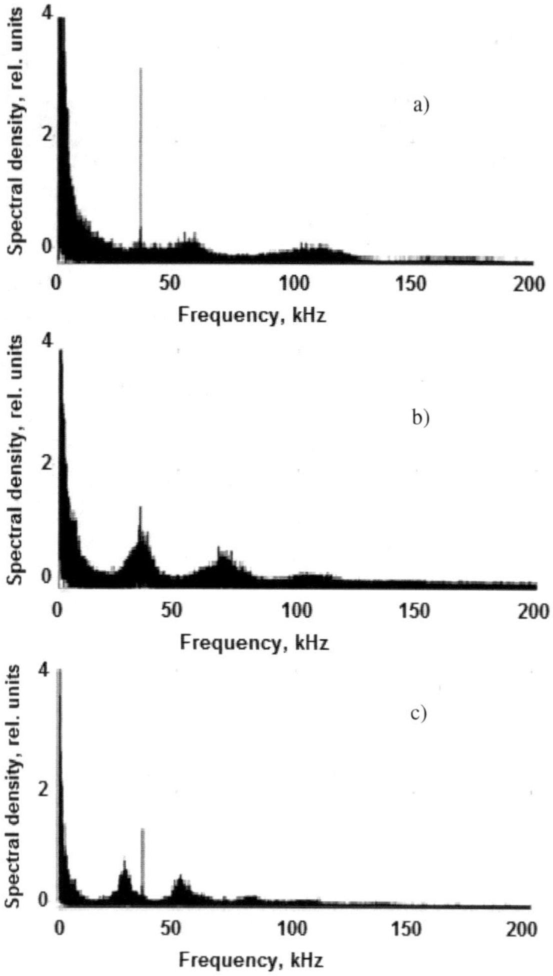

Figure 4. Spectra oscillations secondary emission current in the plasma produced in the laser welding zone in vacuum at different distances Z from the collector of charged particles to the metal surface (a – 10 mm, b – 20 mm, c – 30 mm)

Thus, on the basis of experimental studies of the amplitude-time parameters of current pulses in plasma with pulsed laser welding, the results were obtained. These results will be the basis for the development of methods for the secondary emission control of the process of weld formation during laser welding in vacuum.

Acknowledgments

The reported study was partially supported by the Government of Perm Krai, research project No. C-26/794.

References

[1] Vedenov A A and Gladush G G 1985 *Physical processes in the laser processing of materials.* (Moscow: Energoatomizdat)
[2] Grigoriants A G 1989 *Fundamentals of laser processing of materials* (Moscow: Mechanical

Engineering)
[3] Vinogradov B A, Gavrilenko V N and Libenson M N 1993 *Theoretical bases of influence of laser radiation on materials* (Blagoveshchensk: BPI)
[4] Zuev I V 1997 *Processing materials with concentrated energy flows* (Moscow: MPEI)
[5] Mirkin L I 1975 *Physical bases of processing materials with laser beams* (Moscow: Moscow State University Publishing House)
[6] Letyagin I Yu 2016 Technological perspectives of laser welding in vacuum *Proc. of Int. NTC Welding and related technologies*, pp 6-7
[7] Belenkiy V Ya, Trushnikov D N and Letyagin I Yu 2014 Laser welding in vacuum is a promising welding technology for the production of critical products *Bulletin of the Perm National Research Polytechnic University* (*Mechanical engineering, materials science vol 16*) **chapter 4** pp71-81
[8] Letyagin I Yu, Belenkiy V Ya and Trushnikov D N 2016 Benefits and Prospects of Laser Welding Application in Vacuum *KnE Materials Science* **vol 2016**
[9] Belenkiy V Ya and Trushnikov D N 2013 Investigation of the formation of a secondary current signal in a plasma in electron-beam welding with electron beam oscillation *Welding International* **vol 27** pp 877-880
[10] Boyko V A 1980 *X-ray spectroscopy of laser plasma* (Moscow: Results of science and technology, Radio Engineering)
[11] Trushnikov D N, Belenkiy V Ya and Salomatova E S 2016 Formation of a Secondary Current Signal in Electron Beam welding of Dissimilar Materials *Global Journal of Pure and Applied Mathematics* **vol 12 N** 1. pp 657-676
[12] Belenkiy V Ya and Trushnikov D N 2013 Investigation of the formation of a secondary current signal in a plasma in electron-beam welding with electron beam oscillation *Welding International* **vol 27** pp 877-880
[13] Letyagin I Yu, Trushnikov D N and Belenkiy V Ya 2017 The studies of plasma torch processes by laser beam welding *Materials Science Forum* **vol 893**
[14] Letyagin I Yu, Belenkiy V Y and Trushnikov D N 2018 On the connection between the energy parameters of secondary emission signals from the laser beam welding zone in vacuum with the parameters of metal penetration *Journal of Physics: Conference Series* **T 1109 Issue 1 article 012013**, pp 1-9
[15] Gureev D M and Yamschikov S V 2001 *Fundamentals of the physics of lasers and laser material processing* (Samara: Samara University)

IOP Conf. Series: Materials Science and Engineering **759** (2020) 012016 doi:10.1088/1757-899X/759/1/012016

Express evaluation of welded joints cool resistance

V M Matyunin, A Yu Marchenkov, M A Karimbekov, A N Demidov, P V Volkov, N Abusaif, E I Geraskin and D A Zhgut

National Research University "Moscow Power Engineering Institute", 14, Krasnokazarmennaya str., Moscow, 111250, Russia
E-mail: MatiuninVM@mpei.ru

Abstract. Low-temperature instrumented indentation tests of welded joints are carried out. The effect of metal cooling on indentation diagrams and metal hardness is investigated. Impact strength tests with a gradual decrease in temperature with the identification of the cold brittleness threshold and determination of the critical brittleness temperature for several structural steels are performed. For an express evaluation of a metal cold resistance, the parameter γ is proposed, which is equal to the ratio of hardness at a given low temperature to hardness at room temperature. The relation between the parameter γ and the temperature coefficient of hardness, which is included in the equation describing low-temperature dependence of hardness, is established. For several grades of carbon and alloy steels, the relation between the parameter γ and the critical brittleness temperature $T_{0.4}$, determined by the criterion value of impact strength $KCV = 0.4$ MJ/m^2, is established. For welded joints obtained by electron beam welding and arc welding, the values of $T_{0.4}$ are determined by the parameter γ in various local zones, including parent metal, heat-affected zone and weld metal. The possibility of establishing the acceptable operation temperature of welded joints using the obtained distributions of the critical brittleness temperature $T_{0.4}$ is shown.

1. Introduction

Metal cold resistance characteristics allow establishing the possibility of operating items made of this metal at low temperatures. The evaluation of the metal cold resistance is performed by the critical temperature of brittleness, at which the metal undergoes a transition from a ductile to a brittle state. There are various methods for determining the critical temperature of brittleness [1, 2], which use either values of the mechanical characteristics or fracture analysis data as a criterion of metal brittleness evaluation. To implement these methods, it is necessary to make a test of a series of samples with a gradual decrease in temperature. More often, a series of notched samples is made and tested for impact strength. The temperature dependence of impact strength reveals the cold brittleness threshold of steel, which is used to evaluate the critical temperature of brittleness using the criterion value of impact strength. For typical welded joints, the application of the described methodology is not always possible due to the lack of metal necessary for the manufacture of a standard samples series. Moreover, with the inevitable differences in the structure and mechanical properties of the welded joint local zones (heat-affected zone, fusion zones, weld metal and parent metal), the task of the critical temperature of brittleness evaluation in each zone by the above method is not feasible in practice. Thereby, scientific researches of the cold resistance of welded metal designed for use in the Far North and Siberia are very relevant [3, 4]. The goal of these researches is to develop a method that allows obtaining the distribution of the critical brittleness temperature over the entire welded joint, which would make it possible to determine the upper value of this temperature and to obtain a more reasonable conclusion about the possibility of using the welded construction at low temperatures.

In the work [5], the method was proposed for determining the zero plasticity temperature (T_{ZP}) of metal from the low-temperature change in the values of hardness at the yield strength and hardness at the ultimate strength, determined by indentation. With decreasing test temperature, these hardness characteristics monotonically increase and gradually approach each other in their values. The point of the lines intersection of low-temperature variation of these two hardness characteristics corresponds to the temperature of T_{ZP}. An approximate relationship was revealed between T_{ZP} and the critical brittleness temperature T_{K0} for several grades of carbon steels and pearlitic alloy steels [6]. The temperature T_{K0} was chosen as the higher of two critical temperatures: temperature $T_{0.4}$, obtained by the criterion value of impact strength $KCV = 0.4$ MJ/m^2 and temperature T_{50}, determined by fracture microscopic analysis (T_{50} is the temperature corresponding to a share of ductile fracture of 50% in the fractured surface of the sample). It should be noted that the described method is quite laborious and requires special instruments and devices for determining hardness at the yield strength and hardness at the ultimate strength at low temperatures. This problem can be partly solved by using instrumented indentation with registration of indentation diagrams "load F - displacement α".

In this paper, we developed an express method for the rapid evaluation of the critical brittleness temperature of welded joints local zones by instrumented indentation and suggested informative cold resistance parameters by which it is possible to establish the permissible operation temperature of weld structures in low climatic temperatures.

2. Research methods

A technique of determining hardness using instrumented indentation by a ball indenter, developed at MPEI [7] was applied. According to this technique, using a "F - α" diagram containing loading and unloading curves, it is possible to determine hardness at yield strength and hardness at ultimate without measuring the indent dimensions using optical means. The loading curve of the instrumented indentation diagram "F - α" can be transformed into the diagram "Brinell hardness HB_t - indent depth t/R" [7]. For many structural materials, the HB_t hardness determined in accordance with ISO 6506 is equal to or close to the maximum hardness $(HB_t)_B$. Therefore, it is advisable to study the low-temperature change in HB_t hardness at a constant defined indentation load for the corresponding indenter diameter.

The studies were performed on carbon steels 10 (0.1% C) and 45 (0.45% C) with a change in the test temperature from 293K (20°C) to 193K (-80°C) in increments of 10...20°C. Indentation was carried out using an Instron 5982 universal machine equipped with a chamber where the test metal was cooled by nitrogen vapor. The accuracy of maintaining the low temperature was ± 1°C, the strain rate in compression mode was 0.5 mm/min. A steel ball with a radius of R = 1.25 mm was used as an indenter. According to ISO 6506, the indentation load was 1839 N (187.5 kg).

It is known that in the studied temperature range, the change in HB_t hardness can be quite accurately described by the power equation:

$$HB_t = AT^{-b}, \quad (1)$$

where A and b are constant for a given material.

With deeper cooling of the metal, for example, to a temperature of liquid nitrogen of 77 K (-196°C), it is advisable to describe the temperature dependence of hardness with the equation (2) [8]:

$$HB = HB_{20} + B \cdot \exp(-\beta T). \quad (2)$$

where HB_{20} - Brinell hardness at room temperature; B and β are constant for a given material.

Parameters b and β included in equations (1) and (2) are the temperature hardness coefficients that characterize the metal cold resistance. Moreover, as was shown in [8], the temperature hardness coefficient β correlates with the critical brittleness temperature of low-carbon steels. At the same time, the type of the impact specimen and the criterion value of the impact strength, by which the critical temperature of brittleness were estimated, were not indicated. Thereby, in this research the experimental studies were carried out to establish the relationship between the cold resistance characteristics determined during impact strength tests and the characteristics determined by instrumented indentation. For this, a series of specimens with a sharp notch (Charpy specimens) were prepared to determine the *KCV* impact strength at lower temperatures for several carbon and alloy steels. The tests were carried out on an Instron MPX-450 pendulum equipped with a cooling chamber. The test temperature decreased in increments of 10 degrees from room temperature T_R = 293 K (20°C) to T = 223 K (-70°C). This test temperature range was chosen based on real climatic temperature range in Siberia and the Far North. According to the test results, the temperature dependences of the impact strength *KCV* were obtained with the identification of the cold brittleness threshold and determination of the critical brittleness temperature $T_{0.4}$ by the criterion value of the impact strength KCV = 0.4 MJ/m^2.

Indentation diagrams "F - α" were obtained at T_R = 293 K (20°C) and T = 223 K (-70°C) on the microsections made of these steels, and hardness values at these temperatures $(HB_t)_{TR}$ and $(HB_t)_T$ as well as temperature hardness coefficient b were calculated. The equation for calculating the coefficient b is obtained from equation (1)

$$b = \frac{\ln[(HB)_T / (HB)_{TR}]}{\ln(T_R / T)}. \quad (3)$$

In a given temperature range T_R - T, the coefficient b is related to the ratio $(HB_t)_T/(HB_t)_{TR}$. If this relation is denoted by γ, then, taking into account equation (3), we obtain

$$b = \frac{\ln \gamma}{\ln(T / T_R)}; \quad (4)$$

$$\gamma = \exp[b \cdot \ln(T_R / T)]. \quad (5)$$

3. Results and discussion

In order to study the effect of cooling on the instrumented indentation diagrams "F - α" and "HB_t - t/R", carbon steels "10" and "45" were tested by indentation in the temperature range from T = 293 K (20°C) to 193 K (-80°C). Figure 1a shows the instrumented indentation diagrams "F - α", and in figure 1b shows the "HB_t - t/R" diagrams for the "45" steel at two temperatures: 293 K (20°C) and 193 K (-80°C).

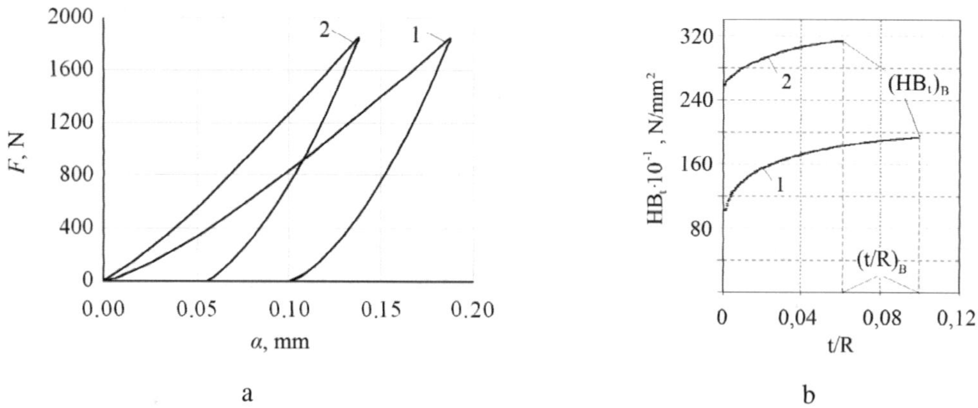

Figure 1. Instrumented indentation diagrams "F - α" (a) and "HB_t - t/R" (b) for steel "45", obtained by a ball indenter D = 2.5 mm, at 293 K (20°C) (1) and 193 K (-80°C) (2).

As it can be seen from figure 1a, the loading curve of the "F - α" diagram rises more abruptly at low temperature, while the full elastoplastic displacement at the maximum indentation load and the residual displacement decrease. These changes are also reflected in the "HB_t - t/R" diagram (figure 1b). At low temperature the curve is located much higher and the maximum value $(HB_t)_B$ is reached at a smaller relative depth $(t/R)_B$. Moreover, the value $(HB_t)_B$ increases significantly.

These experiments made it possible to obtain low-temperature dependences for carbon steels "10" and "45" (figure 2) and to verify the validity of equations (1) - (3).

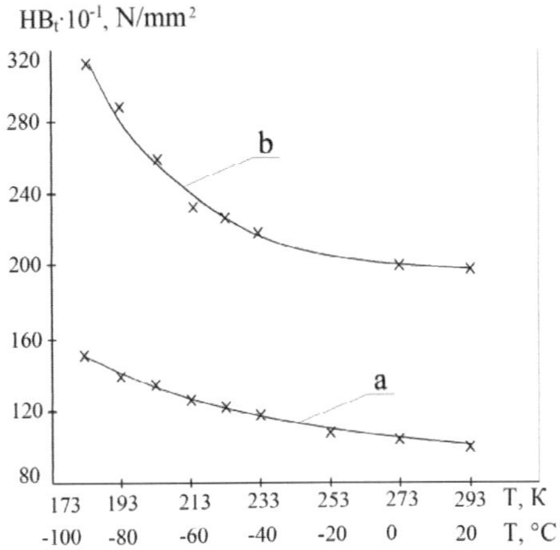

Figure 2. Change in HB_t hardness of steel "10" (a) and steel "45" (b) during cooling.

Impact strength tests with a decrease in temperature were carried out with the same carbon steels, as well as with several alloy steels. According to the results of impact tests of a series of specimens (at least 10 specimens for each steel), low-temperature dependences of impact strength were obtained with the identification of the cold brittleness threshold and determination of the critical brittleness temperature $T_{0.4}$. Figure 3 shows such dependences for carbon steel "10" and alloy steel 12Kh2MFA (0.12%C, 1.5% Cr, 0.4% Mo, 0.2% V). The intersection points of the dashed horizontal line at the level of the criterion value of 0.4 MJ/m^2 correspond to the critical brittleness temperature $T_{0.4}$.

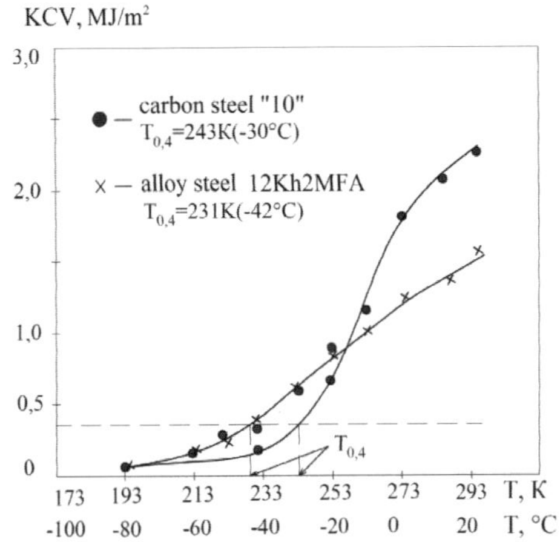

Figure 3. Impact strength KCV of carbon steel "10" and 12Kh2MFA alloy steel at low temperatures.

Then, for all tested steels "F - α" diagrams were obtained at 293 K (20°C) and 203 K (-70°C) and the γ parameter was determined at these temperatures. A correlation relationship was revealed between the critical brittleness temperature $T_{0.4}$ and the cold resistance parameter γ (figure 4).

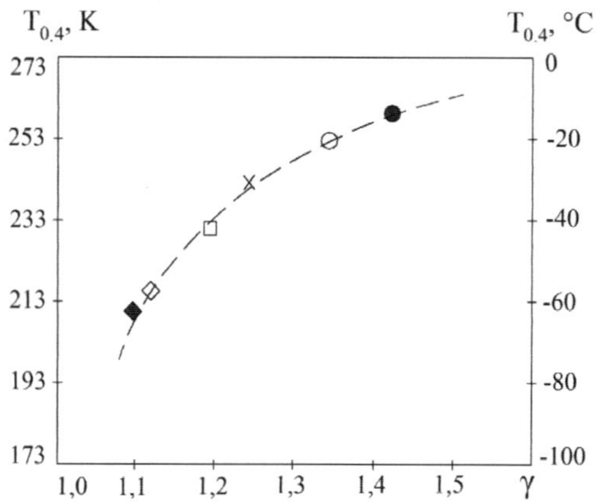

Figure 4. Relationship between the critical brittleness temperature $T_{0.4}$ and the cold resistance parameter γ for steels: ● – carbon steel St0 (≤ 0.23% C, ≤ 0.06% S, ≤ 0.07% P); ○ – carbon steel VSt3sp (0.18% C, ≤ 0.05% S, ≤ 0.04% P); × - steel 10; □ – 12Kh2MFA; ◊ - 15Kh1M1F (0.15% C, 1.2% Cr, 1% Mo, 0.2% V); ♦ - 09G2S (0.09% C, 1.5% Mn, 0.7% Si).

Thus, the easily determined parameter γ also allows quantifying the critical temperature of brittleness during cooling to a given temperature. However, it should be noted that in order to clarify the established relationship between $T_{0.4}$ and γ when cooling the material to -70°C, additional experiments on steels of other grades are required. However, at the same time, the γ parameter already allows lining up several controlled materials by cold resistance.

After establishing the relationship between the cold resistance characteristics determined by instrumented indentation and impact strength tests, experiments were carried out to determine the cold resistance characteristics of the local zones of welded joints according to the technique proposed in the paper. The experiments to determine the hardness HB_t, the cold resistance parameter γ, and the critical brittleness temperature $T_{0.4}$ were carried out on microsections of welded joints obtained by electron beam welding (EBW) and manual arc welding (MMA welding). On each section, at the designated points along the line covering the parent metal, the heat-affected zone and the weld metal, 7-8 instrumented indentation diagrams "F - α" were registered at T = 293 K (20°C), and then at T = 203 K (-70°C). The obtained results made it possible to obtain the distribution of $T_{0.4}$ in various local zones of each welded joint.

In figure 5a there is a graph of the $T_{0.4}$ distribution in various local zones of a welded joint made of steel 38Kh2N2MA (0.38% C, 1.5% Cr, 1.5% Ni, 0.25% Mo) obtained by EBW, and in figure 5b there is a graph for a welded joint of steel 18Kh11MNFB (0.18% C, 11% Cr, 1% Mo, 0.75% Ni, 0.3% V, 0.3% Nb) obtained by MMA welding. Analysis of figure 5a and 5b shows that the welded joint obtained by MMA welding is more cold-resistant than the welded joint obtained by EBW. At the same time, the distribution of $T_{0.4}$ for EBW joint is smoother.

According to the graphs similar to the graphs of figure 5, it is possible to identify the most unfavorable cold resistance local zone for each welded joint and evaluate its critical brittleness temperature. This makes it possible to more reasonably recommend the operating temperature of products with welded joints in conditions of low climatic temperatures.

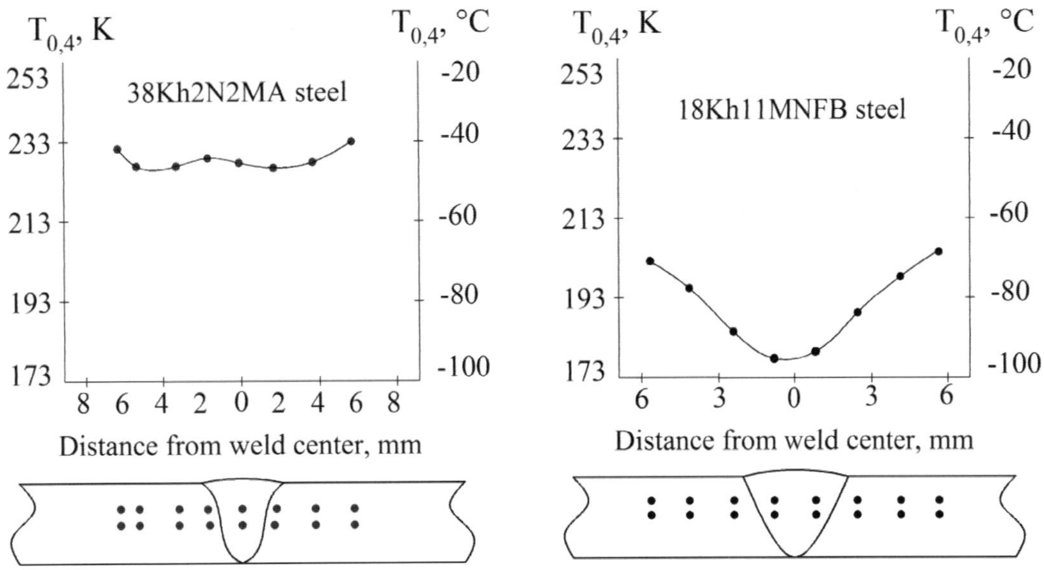

a b

Figure 5. Distribution of the critical brittleness temperature $T_{0.4}$ in the local zones of welded joints obtained by EBW (a) and MMA welding (b)

4. Conclusions

The possibility of an operative evaluation of instrumented indentation of the critical brittleness temperature of local welded joints zones, including parent metal, heat-affected zone and weld metal, is performed. The metal critical brittleness temperature $T_{0.4}$, determined by the criterion value of impact strength $KCV = 0.4$ MJ/m^2, can be estimated by the proposed cold resistance parameter γ, which is equal to the ratio of hardness at a given low temperature to hardness at room temperature. The larger the parameter γ, the higher is the $T_{0.4}$ and lower the cold resistance of the metal. The distribution chart of $T_{0.4}$ in various local zones of the welded joint allows identifying the upper value of $T_{0.4}$ and to recommend the appropriate operating temperature range for the weld joint in conditions of low climatic temperatures.

References

[1] Makhutov N A 1981 *Factory Laboratory* **9** 78–81
[2] Kopelman L A 1978 Resistance of welded joints to brittle fracture (Leningrad: Mashinostroenie) p 232
[3] Efimenko N G, Bartash S N and Sitnikov P A 2018 *Welding production* **11** 16-9
[4] Tarnovsky A I and Geller A B 2018 *Welding production* **2** 13-9
[5] Matyunin V M, Volkov P V , Yudin P N and Poruchikov A V 1988 *Factory Laboratory* **10** 53-6
[6] Matyunin V M 2006 Express diagnostics of structural materials mechanical properties (Moscow: MPEI) p 141
[7] Matyunin V M 2015 Indentation in the diagnosis of the mechanical properties of materials (Moscow: MPEI) p 288
[8] Krivonosova E A 1997 The influence of the coating composition on the quality of the weld metal and the technological properties of the electrode (Yekaterinburg: Perm state tech. un-ty) p 18

IOP Conf. Series: Materials Science and Engineering **759** (2020) 012017 doi:10.1088/1757-899X/759/1/012017

Modification of the surface of steel 3Kh2V8F by application B-Al-coatings by methods of thermal-chemical treatment and electron-beam processing

U L Mishigdorzhiyn[1,3], N S Ulakhanov[1], A S Milonov[2], A P Semenov[2]

[1] East Siberia State University of Technology and Management, 40V Klyuchevskaya street, Ulan-Ude, 670013, Russia
[2] Institute of Physical Material Science of the Siberian Branch of the Russian Academy of Science, 6 Sakhyanovoi street, Ulan-Ude, 670047, Russia
[3] Email: druh@mail.ru

Abstract. The development of new protective coatings is of great fundamental and applied importance for increasing operational properties of surface layers in machine parts, increasing their durability and expanding their functionality. The study is devoted to the creation of coatings based on boron and aluminum on the surface of alloyed steel using a method, combining diffusion saturation (DS) and subsequent electron beam processing (EBP). DS was carried out in saturating pastes based on boron carbide and aluminum at temperature of 1050 °C for 2 hours. As a result of processing, a diffusion layer with thickness of up to (5.6–5.8) × 10^2 μm and complex structure with depth-heterogeneous composition was formed on the steel surface. The subsequent EBP led to a complete transformation of the primary diffusion layer and an increase in its thickness to 10^3 μm. XRD analysis showed significant differences in composition before and after EBP: after EBP tungsten borides (WB, W_2B_9) and iron (Fe_2B) were detected. In addition, it was determined that the distribution of microhardness and elemental composition (B, Al, W) over the depth of the layer after EBP have a more favorable profile without significant fluctuations compared to the sample after DS.

1. Introduction

Currently alloyed and special steels are selected for the production of parts operating under conditions of high mechanical and thermal stresses, accompanied by shock loads. Often the surface layers of these steels need additional hardening due to high mechanical and thermal loads. Thus, it is known that during injection molding, more than 80 % of all dies fail due to the formation of surface cracks caused by thermal stresses [1]. Thus, one of the important challenges in mechanical engineering is the development of methods for improving functional properties of surface layers by various hardening methods.

Improving functional properties of parts made of tool steels can be provided by thermal-chemical treatment (TCT). TCT is of particular interest due to its ability to provide enhanced physical and mechanical properties at low costs. For example, boriding provides a number of positive properties, such as high hardness, low coefficient of friction, high wear resistance, resistance to erosion, corrosion and oxidation [2]. At present, in mechanical engineering technology, when modifying the surface layers of steels and alloys, it is possible to efficiently use multicomponent TCT, which comprises of

IOP Conf. Series: Materials Science and Engineering **759** (2020) 012017 doi:10.1088/1757-899X/759/1/012017

simultaneous or sequential diffusion saturation of the surface of metals and alloys with several chemical elements. Such complex TCT methods as boroaluminizing can significantly increase wear resistance, as well as increase oxidation resistance at high temperatures, corrosion resistance and other properties of the surface layers of machine parts [3].

2. Identifying the problem

Despite a significant number of scientific and technical publications, the problem of finding effective methods of hardening remains relevant. Today, the task of developing a theory applicable to the formation of boroaluminized coatings having a set of physic-mechanical properties — high wear, heat and corrosion resistance — is still not carried out. Conventional boroaluminizing leads to the formation of a layer in which the phases, arranged in layers, do not allow the positive properties of the boroaluminized layer to fully manifest themselves. Existing methods for creating protective coatings usually involve the use - as a basis - of low-cost carbon structural steels subjected to surface modification: the economic effect is achieved by replacing expensive alloy steels and alloys. However, this approach has several limitations:

1. The boron-based coatings obtained by diffusion methods have several disadvantages: the layered structure of the coating, high brittleness, a sharp difference in properties during the transition from the coating to the substrate, the multidirectional nature of the distribution of residual stresses due to the anisotropy of the thermal expansion of boride phases.
2. For tools, working under pressure and under conditions of shock loads and elevated temperatures, there remains a need to ensure strength and high mechanical properties not only on the surface, but also inside their volume. Carbon steels often cannot provide this condition.

In this regard, it is particularly promising to study methods for creating composite structures on the surface of alloyed steels based on boron and aluminum so that both solid and plastic phases are located in the coating in various combinations (ordered or random). It is known that electron beam processing (EBP) of boronized layers leads to phase redistribution and significant microstructural changes [4]. So, as a result of processing, a Fe_3B phase is formed in addition to the existing FeB and Fe_2B. Moreover, instead of the "classic" needle boronized coating, coatings with a composite structure are obtained. In [5] it was established that the wear resistance and ductility of this type of coating is higher than that of the needle layer. Thus, the use of EBP is an effective method of leveling the high fragility of boron-based coatings due to the formation of coatings with a composite structure. This is achieved by phase transformation as a result of the surface layers being re-melted under the influence of an electron beam, which provides a smoother distribution of microhardness, a favorable distribution of residual stresses and improved adhesion properties [6].

The purpose of this work is to obtain a coating with a composite structure to increase the complex of mechanical properties of the surface of 3Kh2V8F steel using the combined method of thermal-chemical and electron-beam treatment.

3. Materials and methods

The coatings were synthesized on the surface of 3Kh2V8F die tool steel in two stages:

1. First, the process of diffusion boroaluminizing was carried out in pastes containing powders of boron carbide, aluminum, and sodium fluoride as an activator. The methodology for the preparation of pastes and the composition of the powder mixture are described in [7, 8].
2. The second stage involved additional treating the obtained diffusion coatings with an electron beam in an electron-beam installation with an EPA-60-04.2 electron gun with a BUEL control unit and a high-voltage rectifier [9]. Heating was carried out by a continuous beam for 2-5 minutes at specific density of $W = 5.7 \times 10^2$ W/mm^2. The residual pressure in the vacuum chamber did not exceed 2×10^{-3} Pa.

The microstructure of the samples was studied using a METAM RV-34 metallographic microscope. Microhardness was determined on a PMT-3M microhardness tester with a load of 0.5 N. X-ray microanalysis was carried out on a JSM-6510LV, JEOL scanning electron microscope with an INCA Energy 350, Oxford Instruments microanalyzer. The phase composition was determined on Phaser 2D, Bruker X-ray diffractometer in copper radiation in an angular interval of 10 ° - 90 ° 2θ.

The studies were carried out at East Siberia State University of Technology and Management (Science Center "Progress"; Department of Mechanical Engineering Technology, Metal Working Machines and Complexes; Department of Metallurgy and Metal Processing Technology) and Institute of Physical Material Science of the Siberian Branch of the Russian Academy of Science (Physical Material Science laboratory).

4. Discussion

As a result of DS, a coating with thickness of up to (5.6–5.8) × 10^2 μm with a complex structure is formed on the steel surface, as shown Figure 1 (a). Depending on the depth, several structural zones in the coating can be distinguished. Porous outer zone 1 consists of large light crystals oriented in the {001} direction. Zone 2 is located below and is characterized by the presence of small or elongated crystals located in the light matrix. Zones 3 and 4 have a cellular structure: the first is with predominant light phases, the latter is with dark, apparently ferritic crystals.

The subsequent EBP led to a complete transformation of the primary diffusion layer and an increase in its thickness to about 10^3 μm, as shown in Figure 1 (b). A homogeneous layer with no visible signs of zoning (layering) was obtained. On the surface of the layer dendritic formations are found.

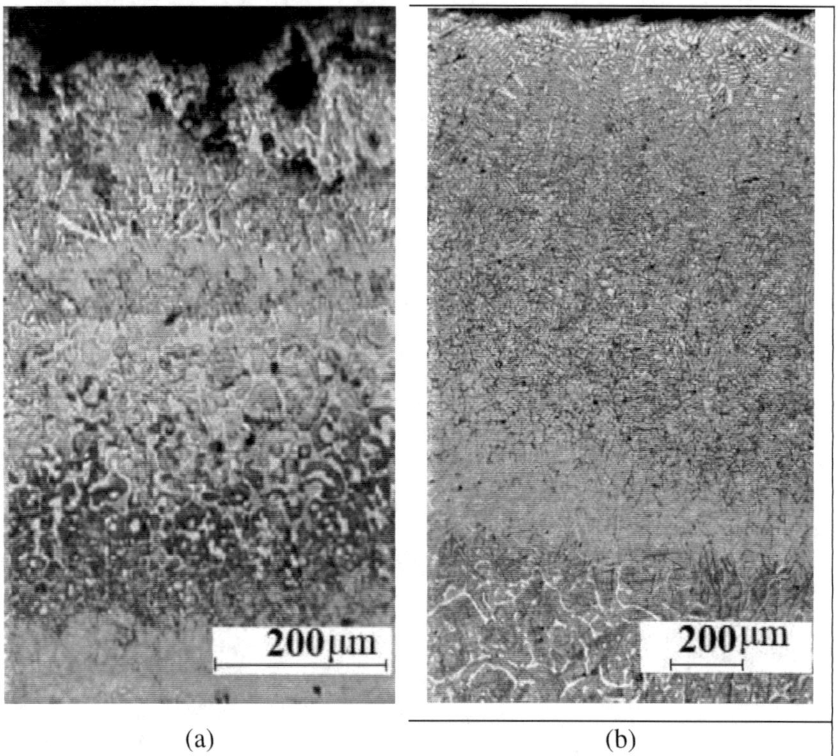

(a) (b)

Figure 1. The microstructure of steel 3Kh2V8F: (a) - after DS; (b) after combined treatment.

XRD analysis revealed the presence of FeB, Fe_3Al, Fe_7W_6, Fe_2O_3 (Figure 2). The presence of iron boride and aluminide is natural for DS in pastes based on boron carbide [7, 8]. The presence of ferrovanadium is unlikely. According to the Fe – W phase diagram, the region of existence of the Fe_7W_6 phase is in the temperature range 1190–1637 °C [10]. The DS process was carried out in a furnace without controlled atmosphere; therefore, the presence of iron oxide can be explained by the diffusion of atmospheric oxygen through the paste.

The electron beam treatment of the diffusion layer led to the formation of vanadium (WB, W_2B_9) and iron (Fe_2B) borides (Figure 3). In this case, FeB boride was not detected. Of the carbide phases, complex carbide Fe_3W_3C-Fe_4W_2C was revealed.

The distribution of microhardness over the depth of the layer showed that the sample after EBP has a more favorable profile without significant fluctuations compared to the sample after MD (Figure 4). The element distribution of boron, aluminum, and tungsten over the depth of the layer after EBP is more uniform compared to the profile after DS (Figure 5, 6).

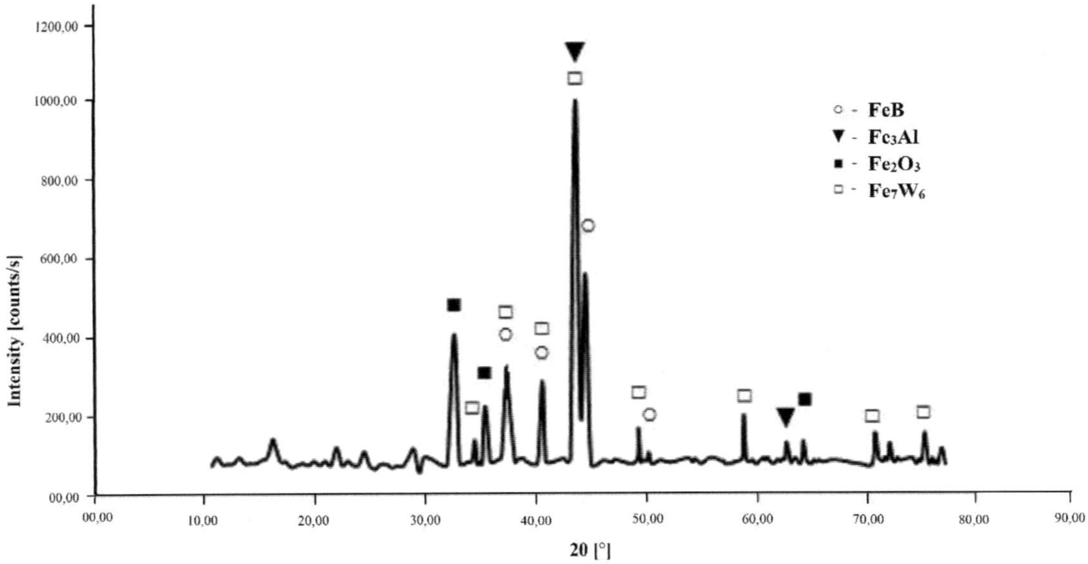

Figure 2. X-ray powder diffraction analysis of steel 3Kh2V8F after DS

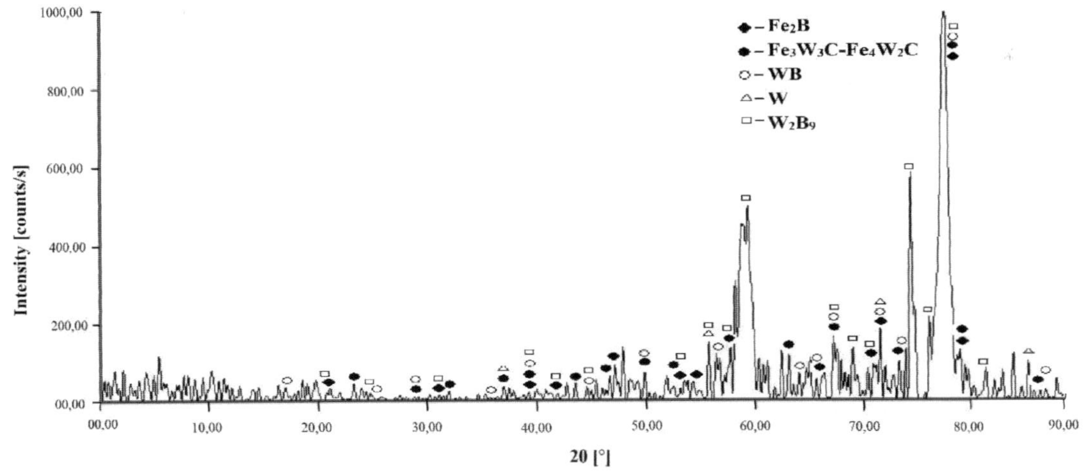

Figure 3. XRD analysis of steel 3Kh2V8F after combined treatment.

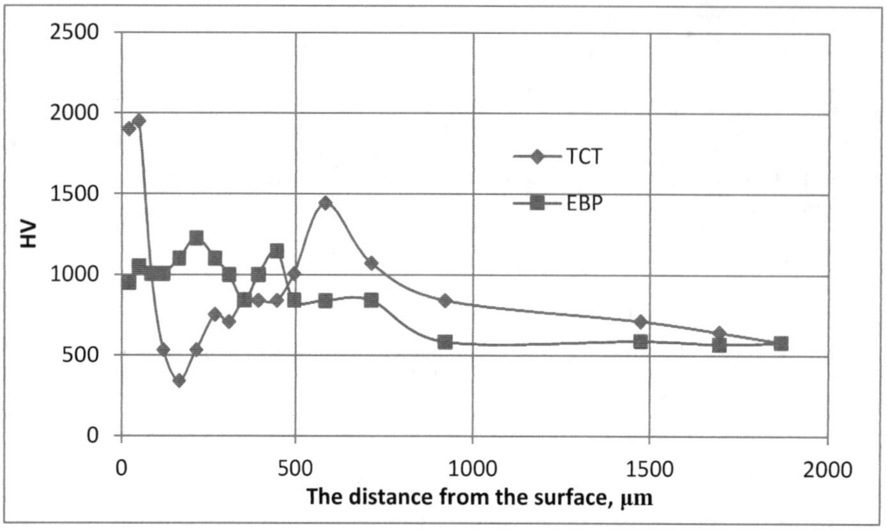

Figure 4. Microhardness distribution of 3Kh2V8F steel after DS (TCT) and combined treatment (EBP).

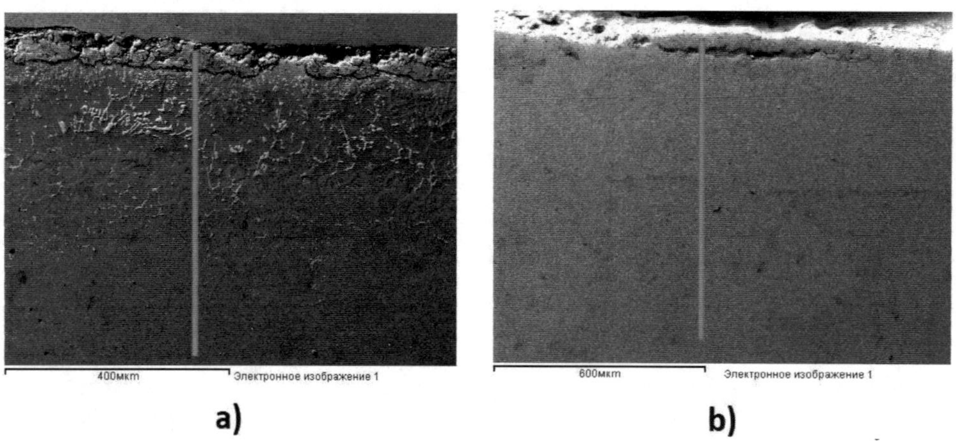

Figure 5. Microstructures of steel 3Kh2V8F with set of spectra lines: (a) after TCT; (b) after combined treatment.

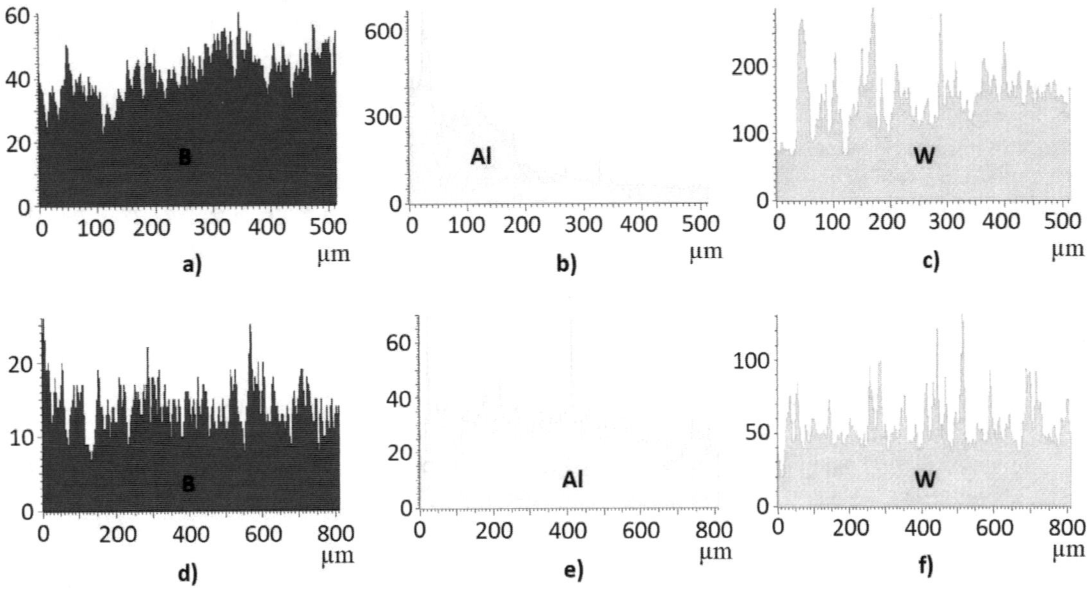

Figure 6. Distribution of elements (B, Al, W) by layer thickness: (a, b, c) - after TCT; (d, e, f) - after combined processing.

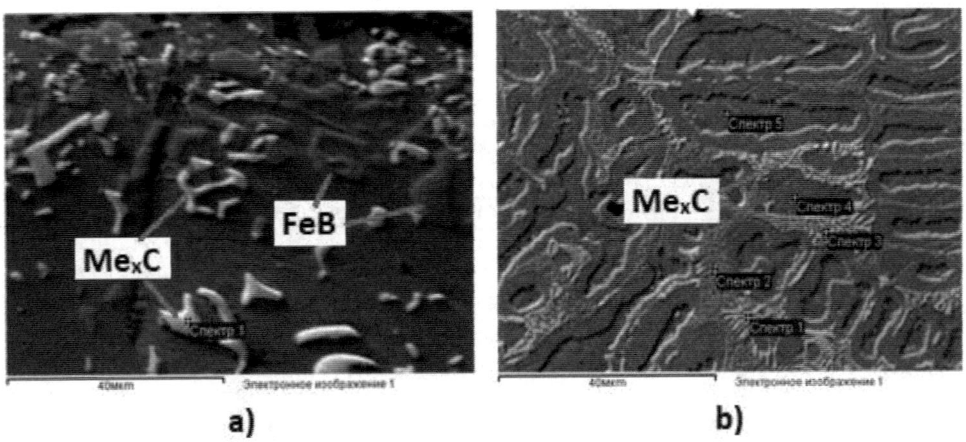

Figure 7. Microstructures of steel 3Kh2V8F, ×1500: (a) – after TCT; (b) – after combined treatment.

SEM images showed that the solid phases - borides and iron carbides differ depending on the treatment by their structure and size (Figure 7).

5. Conclusion

The combined processing method results in equalization of the phase composition along the depth of the layer and obtaining a structure with a uniform composition. Lamination and phase heterogeneity along the coating depth are significantly reduced.

Acknowledgments

The reported study was funded by the grant of the President of the Russian Federation for the state support of young Russian scientists – PhD holders, according to the research project No. 075-15-2019-514.

References

[1] Fuchs K D 2009 Hot-work tool steels with improved properties for die casting applications *6th International tooling conference* pp 183.
[2] Kulka M 2019 Trends in thermochemical techniques of boriding. In Current Trends in Boriding. Engineering Materials *Springer: Cham, Switzerland* 17-98.
[3] Sizov I G, Mishigdorzhiyn U L and Maharov D M 2012 A Study of thermocycling boroaluminizing of carbon steels *Met. Sci. Heat Treat* **53** 592–597
[4] Smirnyagina N N, Semenov A P, Sizov I G, Korobkov N V and Tselovalnikov B I 2000 Electron beam modification and boronation of iron-carbon alloys *Physics and chemistry of materials treatment* **3** 45-49
[5] Smirnyagina N N, Sizov I G, Semenov A P and Vandanov A G 2001Thermodynamic analysis of the synthesis of vanadium borides on the surface of carbon steels in vacuum *Physics and chemistry of materials treatment* **2** 63-67
[6] Sizov I G, Smirnyagina N N and Semenov A P 2001 The structure and properties of boride layers obtained as a result of electron-beam chemical-thermal treatment *Metal Science and Heat Treatment* **11** 45-46
[7] Mishigdorzhiyn U and Sizov I 2018 The influence of boroaluminizing temperature on microstructure and wear resistance in low-carbon steels *Mater Perform Charact* **7** 252–265
[8] Ulakhanov N S, Mishigdorzhiyn U L, Greshilov A D, Tikhonov A G 2019 Technological support for the operational properties of the surface layer of die for hot bending *XII International Conference «Aviamechanical engineering and transport* (Irkutsk: publishing IRNITU) pp 211-217
[9] Grigoriev Yu V, Semenov A P, Narkhinov V P, Gyrylov E I, Druzhinin V V, Kirillov E A and Smirnyagina N N 1992 Powerful melting technological furnace with electron beam heating *Complex use of mineral resources of Transbaikalia* 139-148
[10] Predel B 2013 Fe-W (Iron-Tungsten). In Dy-Er – Fr-Mo. Landolt-Börnstein - Group IV Physical Chemistry (Numerical Data and Functional Relationships in Science and Technology) Madelung O Eds. *Springer: Berlin, Heidelberg, Germany* **5e** 1-4

Influence of technological parameters of spraying on mechanical properties of gas-dynamic coatings

G V Moskvitin, M S Pugachev

Institute of Machine Science named by A.A.Blagonravov, Russian Academy of Sciences, 4 Maly Kharitonyevsky Street, Moscow, 101990, Russia

E-mail: Gvmoskvitin@yandex.ru

Abstract. A methodology for shear and tension "ring tear-off" tests of copper, aluminum, zinc and nickel coatings laid through gas-dynamic spraying has been considered. It has been shown that the shear and tension test methodology permits obtaining dependences of strength of the bond between the coatings and the substrate on their laying temperature. The cohesive strength tension tests of the copper, aluminum, and zinc coatings using the "ring tear-off" method have shown the minimal spread of the values ($\leq$ 5%) at the brittle character of the coating failure, which enhances the plausibility of the obtained mechanical properties.

1. Introduction

By virtue of enhancing the quality of the surface of articles intended for various purposes, by laying various coatings, its resistance to failure under external loads can be increased. When exploiting an article, there always appear loads and their respective stresses, which act along the normal or tangent to the coating surface and result in its failure, the early breakdown of equipment or accidents. In order to avoid this, coatings are tested for their cohesive strength (cohesion) and the strength of the bond between the laid metal layer and the substrate (adhesion).

2. Conducting research and experiments

Copper, aluminum, zinc and nickel coatings were sprayed onto the surface of a 40X steel substrate using a DIMET-404 gas-dynamic installation. The coatings were formed from mechanical mixture of fine powders of metals (Cu, Al, Zn, Ni) and hard brittle particles of corundum (Al_2O_3) [1].

The coatings were laid from a distance of 10 mm from the installation nozzle edge to the substrate surface, the nozzle movement speed was 10 mm/sec. For each powder, a spraying temperature was selected taking into account its physic-mechanical properties and was in the range of 180–540°C.

The value of adhesion between the coating and the substrate was determined using the method of tearing the laid layer from the substrate during a shear test [2]. The coatings were loaded on an INSTRON universal servohydraulic test machine with recording a load-displacement diagram. A sample with a coating was installed in a matrix, and under a load (*P*, figure 1) at a constant movement speed 8×10^{-6} m/sec, it was pushed through until a dramatic drop in the resistance to the load; at that, the coating was displaced relative to the sample or displaced with failure.

IOP Conf. Series: Materials Science and Engineering **759** (2020) 012018 doi:10.1088/1757-899X/759/1/012018

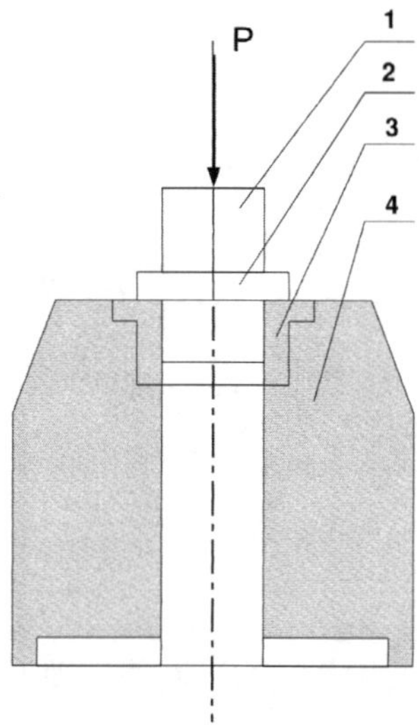

Figure 1. Scheme to test the samples for shear adhesion. Sample — 1; metal layer — 2; matrix — 3; casing — 4.

The limit of strength of the bond between the coating and the substrate was calculated according to the formula (1) [3]:

$$\tau = P(\pi D h)^{-1} \tag{1}$$

,where P is the maximal load at the moment of failure — shear of the coating layer, N; h is the width of the coating belt, m; D is the sample diameter, m.

To test the cohesive strength of the coating, the samples had been made on the basis of an industrial standard that regulates tests of coatings laid using gas thermal methods [4]. A sample consisted of two symmetric parts combined by means of a centering bushing and fastened with a pin and nuts (figure 2).

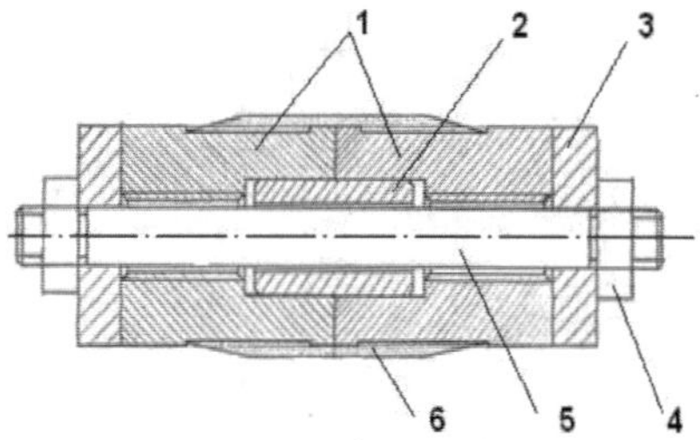

Figure 2. Sample to perform the tension test of the coating using the "ring tear-off" method: 1 — sample parts, 2 — centering bushing, 3 — seal washer, 4 — nut, 5 — pin, 6 — coating.

The coating was laid onto the sample surface and then mechanically treated until the set thickness of the layer was obtained. After that, the pin was removed, and tails were screwed into each part in order to install the sample into the clips of the test machine. The appearance of the coating on the sample surface after the failure (Fig. 6) gave a name to the proposed methodology — a "ring tear-off method" [5].

The samples were tested on a universal servohydraulic testing machine made by SHIMADZU firm, at the maximal load of 50 kN. The load was applied at the constant speed of the actuator's movement equal to 5×10^{-6} m/sec. During the tension test, the data were registered in load-displacement coordinates (figure 3).

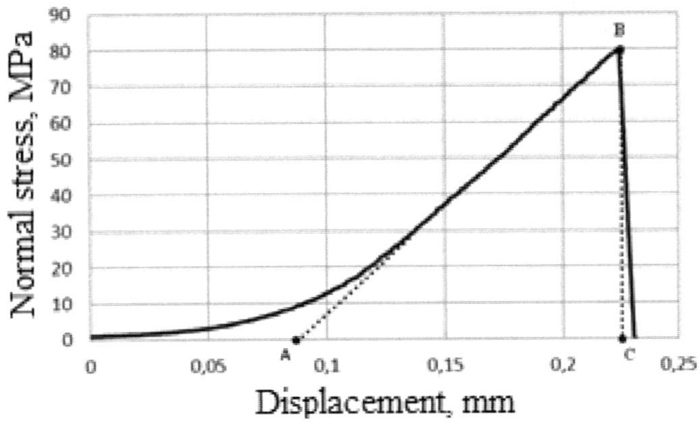

Figure 3. Tension test diagram for the aluminum coating.

The main mechanical characteristics of the coating were determined according to the State Standard GOST 1497-84. The coating's cohesive strength value was calculated in compliance with the following formula (2) [5]:

$$\sigma_{coh} = P_{max} F^{-1} \quad (2)$$

,where P_{max} is the maximal load before the failure, N; F is the coating area, mm^2.

The area of the sprayed metal layer was calculated as the difference between the area of the sample cross-section with and without the coating.

The load rigidity was estimated as the ratio of the limit of the elastic region of the stress-displacement diagram (line segment BC, figure 3) to its respective displacement (line segment AC, figure 3).

3. Research and test results

After the shear tests of the copper coating, the substrate surface contained coating residues, which covered less than 50% of its initial area, which permits interpreting the results as the strength of the bond between the coating and the substrate. A regular texture of red-colored dots can be seen on the substrate surface, and after having sheared the copper coating laid at a temperature of 540°C, on the sample, along the layer borders, there are continuous spots of copper up to 1.8 mm wide, whereon shear strips (figure 4) can be seen. The adhesion between the copper coating and the steel substrate depends on the spraying temperature and achieves the maximal value of 50 MPa at a temperature of 540°C (figure 5).

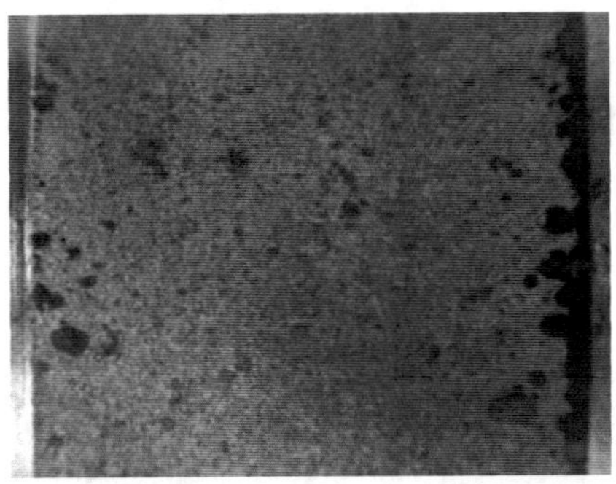

Figure 4. General view of the sample after the shear test

After the shear test, the surfaces of the sample substrates with the laid aluminum and zinc coating have shown presence of the continuous layer of the laid metal on the surfaces across the whole area of the substrate, which permits interpreting the test results as the cohesive strength of the coating after the shear test.

After having been sprayed at an air flow temperature of up to 360°C, the zinc coating's cohesive strength amounts to 35 MPa; an increase of the air flow temperature up to the maximum is accompanied by a decrease in the cohesive strength of the metal down to 14 MPa (figure 5).

The metal's cohesive strength of the aluminum coating does not practically depend on the spraying temperature and amounts to 27 MPa in average (figure 5). The obtained dependence of the cohesive strength on the spraying temperature is rather consistent with the regularities of the changes in the structure and properties of the coating revealed using metallo-physical research methods [6–8].

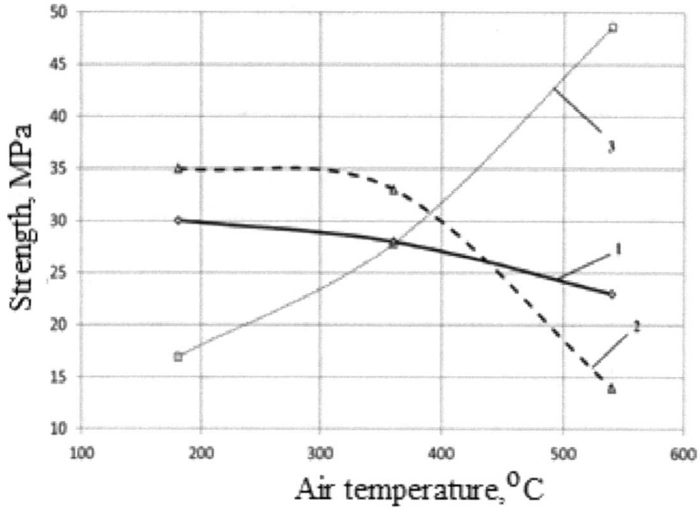

Figure 5. Coating shear test results: aluminum (1), zinc (2) and copper (3).

When considering the results obtained through the shear test of the laid copper, aluminum and zinc coating layer, it should be noted that this method permits revealing the regularities of changes in the strength of the bond between the coating and the substrate as well as of the cohesive strength of the coating metal depending on the technological parameters of spraying. Such tests have to be recommended and used for researches into the mechanical characteristics of coatings laid by means of gas-dynamic spraying.

As a result of the tests of the copper, aluminum, nickel and zinc coatings for their cohesive strength using the "ring tear-off" method, clear dependences of this characteristic on the spraying temperature have been revealed (table 1).

If the spraying temperature rises, the cohesive strength of the aluminum coating is accompanied by a dramatic decrease; at the same time, the coating rigidity decreases and its plasticity grows.

A similar dependence of the cohesive strength on the spraying temperature is traced for the copper coating.

The obtained dependence of the change in the cohesive strength of the copper and aluminum coating on the spraying temperature is rather consistent with the changes in the structure obtained using metallo-physical research methods [9].

The increase in the spraying temperature of the zinc coating does not practically influence the cohesive strength of the coating metal and is accompanied by a decrease in the rigidity and an increase in the plasticity.

Table 1. "Ring tear-off" test results

Metal for spraying	Spraying temperature, °C	Coating strength, MPa	Elongation, mm	Rigidity, MPa/mm
Aluminum	180	181	0.1	1894
	270	108	0.145	746
	360	82	0.144	573
	540	53	0.146	362
Zinc	270	91	0.135	681
	450	95	0.160	594
Nickel	360	64	0.044	1695
	540	72	0.152	473
Copper	180	165	0.091	2375
	270	100	0.134	751
	360	90	0.158	576
	540	86	0.148	585

The rise in the spraying temperature of the nickel coating is accompanied by an increase in the cohesive strength and plasticity, and its rigidity decreases significantly.

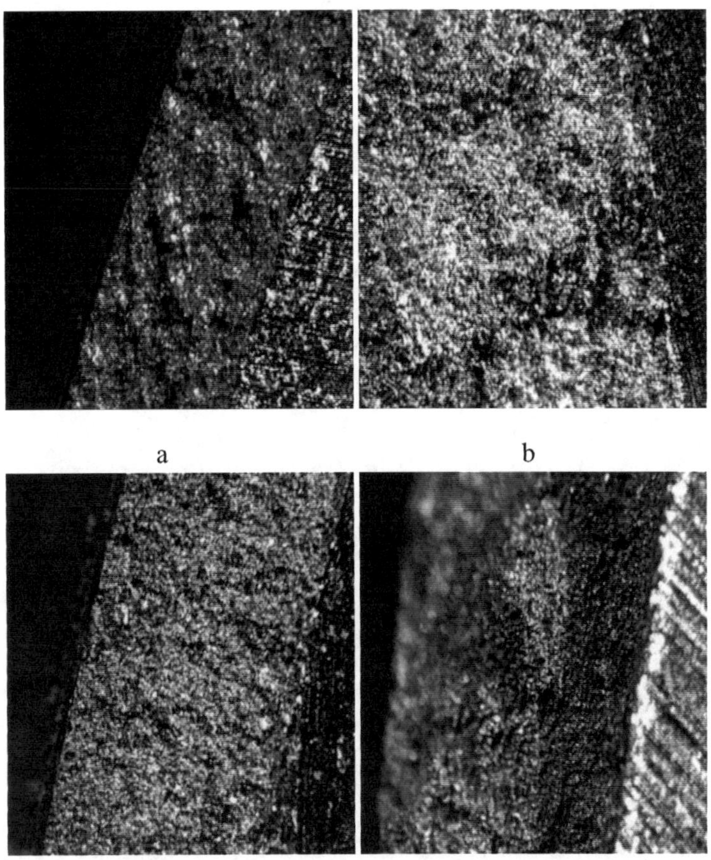

a b

c d

Figure 6. Failure surface of the copper (a), aluminum (b), zinc (c), and nickel (d) coatings.

Both a research of the sample surface after the coating failure and the appearance of the test diagrams show that the failure is brittle (figure 3 and figure 6).

The “ring tear-off” estimate of the coatings’ cohesive strength shows a low spread of the values. The mean deviation of the results in the series of the samples is 0.5 to 5%, which is very important if there is a brittle failure.

4. Conclusions

As a result of the shear test, the determination of the coating’s mechanical properties has permitted obtaining the dependence of strength of the bond between the coating and the substrate on the spraying temperature; at that, if the coating’s adhesion exceeds the cohesive strength of the laid metal layer, the methodology permits estimating the cohesive strength through a shear test.

As the spraying temperature rises, the adhesion between the copper coating and the steel substrate increases and achieves the maximal value of 50 MPa at a temperature of 540°C. The maximal cohesive strength of the aluminum- and zinc-based coatings amounts to 30 and 35 MPa respectively and decreases, as the spraying temperature rises.

The “ring tear-off” determination of the coating’s mechanical properties permits researching their dependence on the technological parameters of spraying.

The maximal values of the coatings’ cohesive strength have amounted to: 181 MPa for aluminum, 165 MPa for copper, 72 MPa for nickel, 95 MPa for zinc; the strength of the aluminum and copper coatings decreases, as the coating laying temperature rises; on the contrary, the strength increases for nickel and does not change for zinc.

The “ring tear-off” method provides a highly credible estimate of the coatings’ cohesive strength.

References

[1] Dimet. Primenenie tekhnologii i oborudovaniya [Dimet. Application of technology and equipment] (electronic resource) Available at: http://www.dimet-r.narod.ru/ (accessed 30 June 2019)

[2] Methodological manual 1987 *Opredelenie prochnosti scepleniya gazotermicheskih pokrytij s osnovoj* [Determination of adhesion strength of thermal coatings with the base] (Moscow: VNIINMASH) [In Russian]

[3] Arkhipov V E, Landarski A F, Melshanov A F, Moskvitin G V, Pugachev M S 2011 Svojstva mednyh pokrytij, nanesyonnyh gazodinamicheskim napyleniem [Properties of copper coatings applied by gas-dynamic spraying] *Uprochnyayushchie tekhnologii i pokrytiya [Hardening technologies and coatings]* **9** 17-23 [In Russian]

[4] Xasui A, Morigaki O 1985 *Naplavka i napylenie* [Surfacing and spraying] Gerevod s yaponskogo V.N. Popova pod red. V.S. Stepina, N.G. Shesterkina [translation from Japanese by V. N. Popov edited by V.S. Stepin, N.G. Shesterkin] p 240 [In Russian]

[5] Arkhipov V E, Landarski A F, Moskvitin G V, Pugachev M S 2019 Ocenka kogezionnoj prochnosti gazodinamicheskih pokrytij [Assessment of cohesive strength of gas-dynamic coatings] *Vestnik mashinostroeniya [Bulletin of mechanical engineering]* **4** 64-70 [In Russian]

[6] Arkhipov V E, Landarski A F, Melshanov A F, Moskvitin G V, Pugachev M S 2015 Tekhnologicheskie osobennosti gazodinamicheskogo naneseniya pokrytij [Technological features of gas-dynamic coating] *Vestnik mashinostroeniya [Bulletin of mechanical engineering]* **9** 64-70 [In Russian]

[7] Arkhipov V E, Landarski A F, Moskvitin G V, Pugachev M S, Falaleev N S 2016 Svojstva alyuminij – cinkovogo pokrytiya, nanesyonnogo gazodinamicheskim napyleniem [Properties of aluminum-zinc coating applied by gas-dynamic spraying] *Uprochnyayushchie tekhnologii i pokrytiya [Hardening technologies and coatings]* **6** 28-34 [In Russian]

[8] Arkhipov V E, Dubrovina A A, Kuksenova L I, Landarski A F, Moskvitin G V, Pugachev M S 2015 Struktura i svojstva pokrytij, nanesennyh gazodinamicheskim napyleniem [Structure and properties of coatings deposited gas-dynamic spraying] *Uprochnyayushchie tekhnologii i pokrytiya [Hardening technologies and coatings]* **4** 18-24 [In Russian]
[9] Arkhipov V E, Landarski A F, Moskvitin G V, Pugachev M S 2017 *Gazodinamicheskoe napylenie: struktura i svojstva pokrytij* [Gas-dynamic spraying: structure and properties of coatings] (Moscow: KRASAND) p 240 [In Russian]

IOP Conf. Series: Materials Science and Engineering **759** (2020) 012019 doi:10.1088/1757-899X/759/1/012019

Effect of molybdenum on the hot-tearing susceptibility of the Ni-Cr-Mo-B system

A S Nazarko[1], R L Plomoyalo[1], D V Dmitrenko[1]

[1] Kuban State Technological University, 2, Moskovskaya Street, Krasnodar, 350072, Russia

E-mail: nazarkoaleksandr@rambler.ru

Abstract. The paper presents data on the effect of molybdenum on the hot-tearing susceptibility of the Ni-Cr-Mo-B system when surfacing with nickel-based alloys. With a boron content of 2.5 – 3.2%, as well as low carbon and silicon content, the alloying elements, depending on a doping concentration, have different effects on the technological strength of the deposited metal. Molybdenum increases the technological strength of alloys. Alloying the deposited metal with molybdenum in an amount up to 12% in alloys with 2.5% boron concentration at 20% chromium does not lead to the disappearance of cracks. At a molybdenum concentration of more than 12%, the number of cracks in the deposited bead decreases and when the molybdenum concentration is equal to 15%, the cracks disappear, and only increasing the molybdenum content to 20%, in the presence of at least 20% chromium in the alloy, makes it possible to obtain deposited metal without cracks.

1. Introduction

Currently, extreme cases of intercrystalline rupture at high temperatures in nonequilibrium solidifying alloys (deposited layers) are clearly distinguished.

1. Solidifying cracks are formed along the enriched interaxial and interdendritic areas at the moment when the latter are in a liquid state. This type of destruction is related to the solidification of the last portions of the liquid phase enriched in easily fusible alloying elements and impurities, and is observed in alloys in which the second phase in the form of eutectics and chemical compounds is formed directly in the process of solidification [1].

2. Subsolidus cracks which form most often in single-phase alloys with a narrow temperature range of solidification not along liquid intercrystalline layers, but after solidification of the last portions of the liquid phase, i.e. below the temperature of the real solidus.

As it is shown by recent researches, the fragmentation of the columnar structure and the formation of a boundary network arbitrarily oriented towards the primary crystallization forms (cells and dendrites) are determined by the subsolidus migration of grain boundaries arisen in solidifying to new, more equilibrium positions with a lower boundary energy. Based on the research findings by M. Kh. Shorshorov [5], hot tearing in nickel-chrome-tungsten alloys occurs in the temperature range (from solidus to 1273 – 1253 K (1000 – 980 °C)), in which polygonization is most likely to develop. In this temperature range there is observed a plasticity drop of alloys. The lowest plasticity occurs at temperatures between solidus and 1573 K and is equal to 0.5 – 0.6 %.

2. Materials and methods

In order to determine the effect of alloying elements on hot-tearing susceptibility, experimental surfacing was carried out with cast rods of a 5 mm diameter and a 270 mm length on 60x15x5 mm plates in 1, 2 and 3 layers. The cast rods [4] of the experimental alloys were melted in an LSZ-35 kW high-frequency induction furnace. The crucible material was chrome magnesite. The melting was carried out using industrial-grade constituents. Primary nickel of the H-2 grade (GOST 849-2008), ferromolybdenum of the FMo60 grade (GOST 4759-91), ferrochrome of the FH005A grade (GOST 4757-91), ferroboron of the FB20 grade (GOST 14848-69) were used as a charge mixture. Mishmetal (ferrocerium) of the MTS50ZHZ grade (Technical Specifications 48-4-280-91) was used as a deoxidizer. The molten metal was poured into green sand molds. The surfacing was performed using an Aurora INTER TIG 200 AC/DC Pulse (TIG, MMA) welding machine. To reduce the process of mixing the deposited metal with the parent one, when surfacing the first layer, the welding current was equal to 100 – 110 A, the second to 115 – 120 A and the third to 130 – 135 A.

3. Effect of molybdenum on hot-tearing susceptibility in surfacing

To select wear-resistant surfacing alloys [2, 3] of the Ni-Cr-Mo-B system without hot cracks, experimental surfacing was performed in a wide range of alloying with the principal constituents — chrome, molybdenum, boron, and the additional ones — vanadium and niobium. The amount of boron in the alloy varied from 2.5 to 4%, based on the experience of alloying surfacing nickel alloys with boron, described in [6, 7]. Carbon content in all surfacing did not exceed 0.05%, thereby reducing the likelihood of carbide formation. The constitutions of the alloys investigated and the number of detected cracks are presented in Table 1. Based on the results obtained (shown in the table below), we derived the molybdenum content dependence of the Ni-Cr-Mo-B alloy cracking (see Fig. 1 – 5).

Table 1. Constitutions of the experimental Ni-Cr-Mo-B alloys

S. No.	Mo, %	Cr, %	B, %	V, %	Nb, %	The number of cracks, pcs	Note
1	5.0	20.0	2.5–3.2	-	-	8–12	1. All alloys shown in the table are nickel based
2	7.0	20.0	2.5–3.2	-	-	6–11	
3	10.0	20.0	2.5–3.2	-	-	5–9	
4	12.0	20.0	2.5	-	-	1	2. The carbon content in all alloys does not exceed 0.05%
5	12,0	20.0	2.8	-	-	2	
6	12.0	20.0	3.0–3.2	-	-	3	
7	15.0	20.0	2.5	-	-	-	3. The iron content in all alloys is at the level of 4%, the manganese 0.8 – 1 %, the silicon 0.12 – 0,2 %
8	15.0	20.0	2.8	-	-	2	
9	15.0	20.0	3.0–3.2	-	-	3	
10	17.0	20.0	2.5–3.0	-	-	-	
11	17.0	20.0	3.2	-	-	1	
12	20.0	20.0	2.5–3.2	-	-	-	
13	20.0	20.0	2.5–3.2	0.6–2.0	-	4–5	
14	20.0	20.0	2.5–3.2	-	1.0–2.0	7–9	

It is known that with a molybdenum content increase, the resistance of nickel-chrome deposited metal to hot tearing gradually increases, reaching a maximum at 22 – 25 %, then rapidly decreases [8].

Many researchers have tried to explain the effect of molybdenum in terms of the polygonizational hypothesis of hot tearing in single-phase alloys at high temperatures. According to this hypothesis, molybdenum, having a high diffusion activation energy, inhibits the movement and grouping of lattice imperfection in a nickel-chromium matrix and inhibits the formation of secondary polygonizational boundaries which are the nuclei of intercrystalline rupture. However, there exists an opinion [9] that, depending on the concentration of particular elements in nickel-based alloys, both subsolidus and crystallization patterns of hot tearing are possible.

According to the literature data [8], with molybdenum content up to 10% in a nickel-chrome alloy, cracks form in solid-liquid state. The molybdenum content increase of more than 10% results in crack formation in the metal already in solid state [10]. However, the effect of molybdenum on the technological strength of hard surfacing nickel-chromium alloys with boron has not yet been fully studied and it is therefore of interest to trace the effect of the concentration of molybdenum and boron on the technological strength of Ni-Cr-Mo-B alloys. As it is shown in Table 1, the addition of up to 12% molybdenum in the alloys with 2.5 – 3.2% boron concentration at 20% chromium does not lead to the disappearance of cracks, however, when comparing alloys containing molybdenum of 5, 7, 10 and 12% the number of cracks in the weld bead decreases and when the molybdenum concentration is 15% the cracks disappear. However, if we follow the concentration of molybdenum and boron further in Table 1, we can note that only molybdenum content increasing to 20% can guarantee obtaining deposited metal without cracks in the most used range of boron doping from 2.5 to 3.2 %. A further increase in molybdenum content in the surfacing metal with an increase in boron concentration of more than 3.2% does not have a positive effect on the technological strength of the deposited metal, since in this case a large number of borides and eutectics are formed, which adversely affects the ductility of the alloys. Therefore, it can be said that molybdenum alloying of nickel-chromium alloys with boron up to 20% increases the technological strength of the deposited metal with boron content up to 3.2%.

Figures 1 – 5 show the dependences of the molybdenum content effect on the susceptibility of the Ni-Cr-Mo-B alloys to hot tearing.

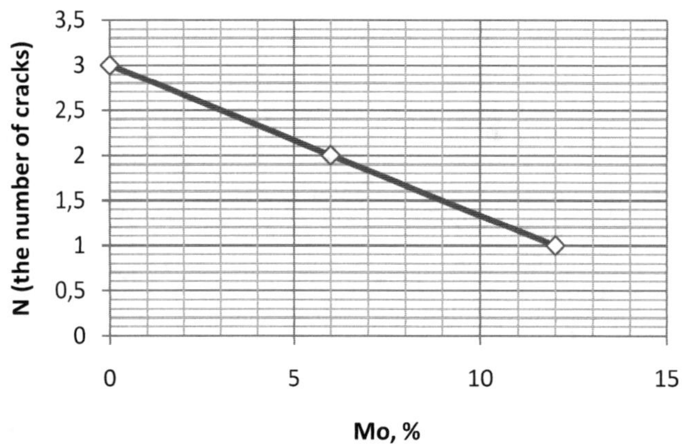

Figure 1. Dependence of crack formation of the Ni-Cr-Mo-B alloys on the molybdenum content (2.8 – 3.2% boron, 20% chromium, and 12% molybdenum).

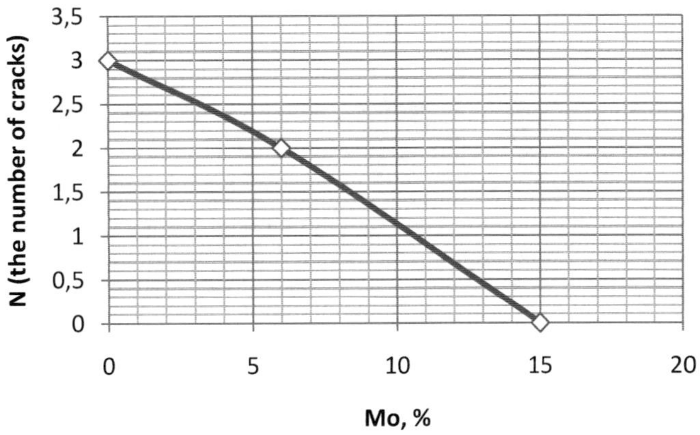

Figure 2. Dependence of crack formation of the Ni-Cr-Mo-B alloys on the molybdenum content (2.8 – 3.2% boron, 20% chromium, and 15% molybdenum).

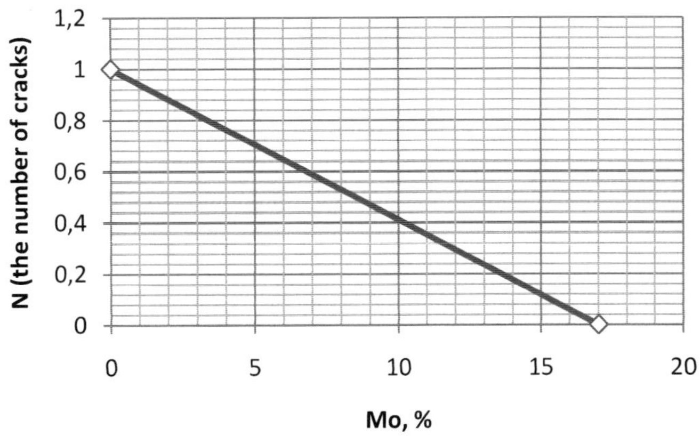

Figure 3. Dependence of crack formation of the Ni-Cr-Mo-B alloys on the molybdenum content (2.8 – 3.2% boron, 20% chromium, and 17% molybdenum).

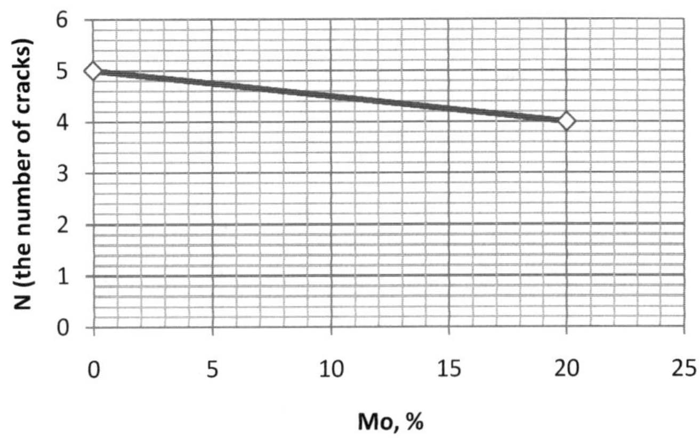

Figure 4. Dependence of crack formation of the Ni-Cr-Mo-B alloys on the molybdenum content (2.8 – 3.2% boron, 20% chromium, 0.6 – 2.0% vanadium, and 20% molybdenum).

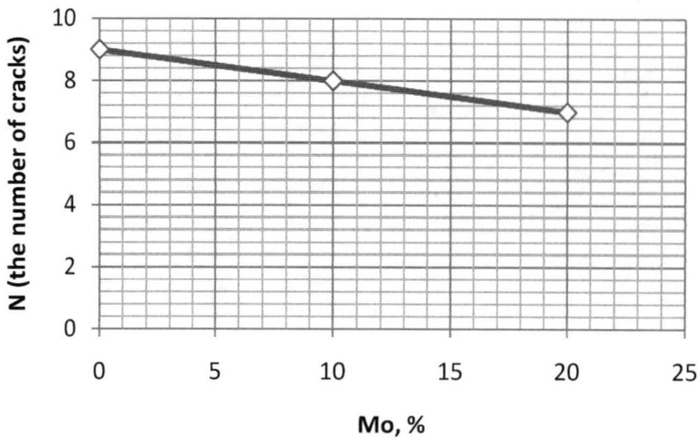

Figure 5. Dependence of crack formation of the Ni-Cr-Mo-B alloys on the molybdenum content (2.8 – 3.2% boron, 20% chromium, 1.0 – 2.0% niobium, and 20% molybdenum).

4. Conclusion

Based on the experimental results obtained, it can be concluded that in nickel-based alloys of the Ni-Cr-Mo-B system with 2.8 – 3.2% boron and a low content of carbon and silicon, molybdenum has different effects on the technological strength of the deposited metal.

Molybdenum improves the technological strength of alloys. Alloying deposited metal with molybdenum in an amount not less than 20% (in the presence of at least 20% chromium in the alloy) makes it possible to obtain the deposited metal without cracks.

Based on the above, it can be concluded that alloying Ni-Cr-Mo-B alloys within: C≤0.05 %, Si≤0.2 %, Mn=0.8 – 1 %, Fe≤4 %, Cr=20 – 30 %, Mo=20 – 25 %, B=2.5 – 3,2 %, Ni – all the rest of it, does not cause hot tearing in surfacing.

References

[1] Movchan B A 1962 Interrelation of physical microinhomogeneity with hot cracks in welded seams *Welding production* **4** 6–8

[2] Nazarko A S, Plomodyalo R L 2019 Retailoring of the bearing surfaces of machine parts by wear-resistant surfacing materials. *In Proc: Engineering Technology of Machine Building. Proceedings of the VIII International Scientific and Technical Conference.* (Omsk. p. 9–14)

[3] Nazarko A S, Plomodyalo R L 2019 Surfacing of the bearing surfaces of machine parts to increase wear resistance in retailoring. *In the collected book: Research and Disign of Intelligent Systems in the Automotive Industry, Aircraft Industry and Machine Building (ISMCA 2019). Proceedings of the III All-Russian Scientific and Practical Conference with International Participation.* (Taganrog, p. 34–39)

[4] Nazarko A S, Plomodyalo R L, Bondar D Yu 2018 Development of the cast rod for surfacing valves. *In Proc: Mechanics, equipment, materials and technologies. Collection of scientific articles based on the proceedings of the International Scientific and Practical Conference dedicated to the 100th anniversary of the Kuban State Technological University.* (Krasnodar, p. 360–365)

[5] Shorshorov M Kh 1961 Metallurgical ways to prevent hot cracks during automatic submerged-arc welding of heat-resistant nickel alloys. *Collection of scientific papers on the theory of high-temperature strength.* (Institute of Metallurgy, USSR Academy of Sciences, p. 173–177).

[6] Gladkiy P V et al. 1968 Plasma surfacing of chromium-nickel alloys doped with silicon and boron *Automatic welding* **9** 58–62
[7] Mnushkin O S 1973 *Structure and properties of Ni-Cr-Si-B alloys for sealing surface surfacing. Collection of scientific papers "Advanced welding methods, new materials and designs in welding industry."* (Izhevsk, p. 89–96)
[8] Loseva G E et al. 1973 Effect of molybdenum alloying of a nickel-chromium alloy on its resistance to hot tearing *Welding Production* **3** 10–12
[9] Shorshorov M Kh 1973 *Hot cracks in welding of heat-resistant alloys* (Moscow: Machine Building)
[10] Sorokin L I et al. 1969 Effect of molybdenum and tungsten on the properties of deposited metal when alloying by coating *Welding Production* **12** 8

IOP Conf. Series: Materials Science and Engineering **759** (2020) 012020 doi:10.1088/1757-899X/759/1/012020

Development of methods for forecasting the microstructure of the welded joint with reference to the temperature-time modes of electron-beam welding

T V Olshanskaya [1,2], **E M Fedoseeva** [1]

[1] Department of Welding Production, Metrology and Materials Technology, Perm National Research Polytechnic University, 29, Komsomolsky Avenue, Perm, 614990, Russian Federation
[2] Correspondent author: tvo66@mail.ru

Abstract. The research is devoted to the development of methods for forecasting microstructure of a welded joint in relation to the temperature-time modes of an electron-beam welding (EBW). The analysis of the simulation models and methods existing in the practice of thermal treatment and relating to austenite decomposition kinetics during cooling was carried out to choose methods for forecasting the forming microstructure in a welded joint at EBW. On the basis of these models one method for forecasting the microstructure of welded joints at EBW is offered in this work.

1. Introduction

An electron-beam welding is considered to be one of the advanced ways of the production of welded joints. The weld formation is essentially dependent on the forming microstructure, being one of the main forecasted joint quality criteria. To forecast the microstructure of a weld metal and a heat-affected zone it is necessary to know thermal cycle for a given section of a welded joint and the criteria by which means the percentage ratio of structural components is estimated.

Generally the criteria for forecasting metal microstructure of a weld and a heat-affected zone are cooling time within a temperature range of 800-200 °C ($t_{8/2}$) and cooling rate within a range of 600-500 °C ($w_{5/6}$), which are determined with the use of thermal welding cycle for the analyzed area. However the heating and cooling rates vary in time nonlinearly during welding, so instantaneous process rates should be applied.

2. Problem identification, materials and methods

The equations for determining the instantaneous heating and cooling rates are derived from thermal problems equations [1]:

$$W(x,y,z,\tau) = \frac{dT(x,y,z,\tau)}{dt} \quad at \quad dt = \frac{dx}{V} \quad W(x,y,z,\tau) = \frac{dT(x,y,z,\tau)}{dx}V.$$

In the planned research the thermal cycles and instantaneous heating and cooling rates were calculated and their analysis was taken to choose the basic criteria for forecasting microstructure formed at EBW. The calculation and the analysis were carried out with the use of the welded joint of the steel 20Cr3MoNbW produced by the serial factory technology (figure 1 and figure 2).

IOP Conf. Series: Materials Science and Engineering **759** (2020) 012020 doi:10.1088/1757-899X/759/1/012020

3. Discussion of results

The comparative analysis of mentioned thermal cycles ($t_{8/2}$, $w_{6/5}$, max $W_{cooling}$) calculated for different areas of the welded joint showed the following information. The cooling time within the temperature range of 800-200 °C ($t_{8/2}$) differs by tenths of a second and the cooling rate at the temperature 550°C ($w_{5/6}$) varies up to 3 °C/s for the weld areas having the same temperature and being located at different depths (table 1).

A similar trend is observed when comparing the data for the weld metal, overheating zone and complete recrystallization that are located at the same depth of the welded joint. In accordance with the obtained $t_{8/2}$ and $w_{6/5}$ data the process of the weld cooling, which is below the temperature of the phase transformation Ae_3, occurs under the same conditions, and therefore the structure in the weld metal and in the heat-affected zone throughout the depth of the welded joint should be the same. However, the metallographic analysis showed that different microstructure was formed throughout the depth of the weld metal (figure 3). An almost martensitic structure with small bainite precipitates is formed in the lower area of the weld, more bainite is located in the upper area of the weld and it precipitates mainly along the boundaries of primary dendritic grains.

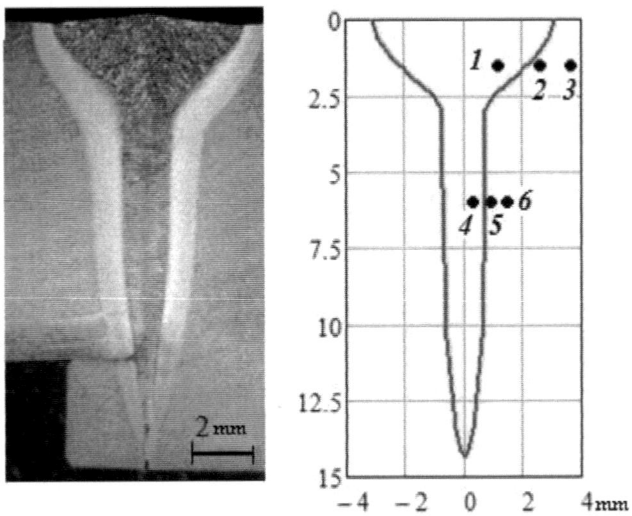

Figure 1. Macrostructure of the welded joint and design shape of the weld; steel 20Cr3MoNbW: I_b = 100 mA; U – 30 кV, V_w – 10 m/hr, points 1,2,3 at h = 1,5 mm; points 4,5,6 at h = 6 mm, 1, 4 – weld metal; 2, 5 – overheating zone; 3, 6 – zone of complete recrystallization.

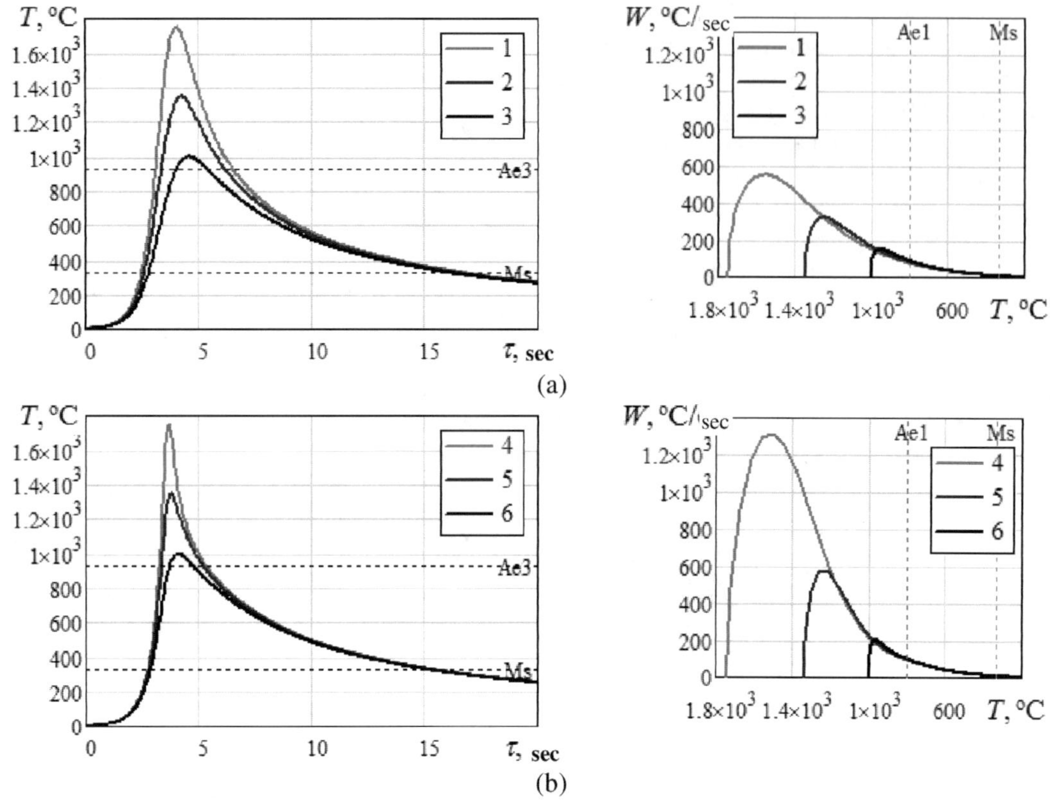

Figure 2. Calculated thermal welding cycles (mathematical dependencies T/τ) and instantaneous cooling rates (mathematical dependencies W/T) ; (a) - at depth h = 1.5 mm; (b) - at depth h = 6 mm. Curves 1, 4 - weld metal; 2, 5 - overheating zone; 3, 6 - zone of complete recrystallization.

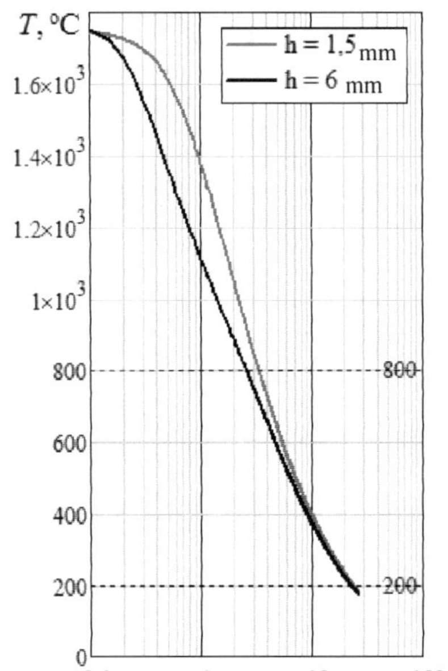

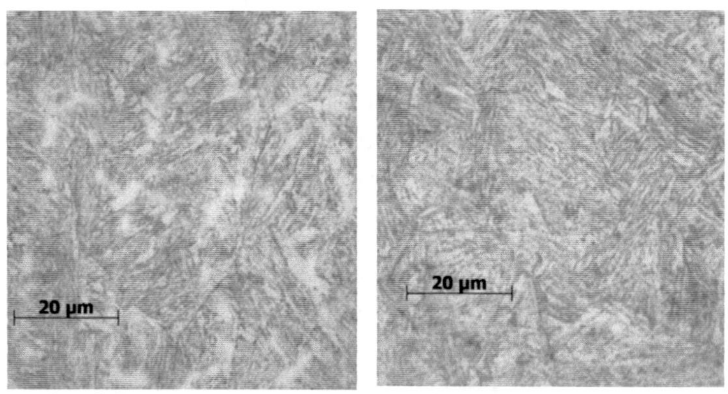

h = 1,5 mm h = 6 mm

Figure 3. Cooling curves from 1750 °C temperature and weld metal microstructure for different depths (Krupp reagent).

Table 1. Parameters of the cooling process for the welded joint, the steel 20Cr3MoNbW.

Cooling parameters	Analyzed weld sections					
	Weld metal		Overheating zone, $T = 1350$°C		Zone of complete recrystallization, $T_H = 1000$°C	
	h = 1,5mm	h = 6mm	h = 1,5mm	h = 6mm	h = 1,5mm	h = 6mm
$t_{8/2}$, s	20,2	19,7	20	19,7	20,3	19,8
$w_{5/6}$, °C/s	26	30	28,5	30,4	29	31
max $W_{cooling}$, °C/s	414	1000	321	564	153	202

Significant differences in thermal cycles and cooling rates are observed in the high-temperature region at $T > 800$ °C (figure 3). The comparative analysis of maximum values of an instantaneous cooling rate (max $W_{cooling}$) showed that maximum values $W_{cooling}$ vary from each other in different areas of the welded joint. At the same time, an increase of max $W_{cooling}$ is observed in the depth of a weld penetration: it increases more than twice in the metal of the welded joint, it rises more than 1,7 times in the overheating zone, and it rises more than 1,5 times in the zone of complete recrystallization (table 1). The values max $W_{cooling}$ decrease along the width of the welded joint.

When forecasting the microstructure of a welded joint metal and a heat-affected zone forming at EBW the criteria should cover the entire cooling history, starting with the maximum temperature reached during heating. Such criteria can be a crystallization temperature and an instantaneous cooling rate at this temperature for a weld metal, and maximum heating temperature and the highest instantaneous cooling rate, in relation to a given thermal cycle, for a heat-affected zone.

The analysis of existing in the practice of heat treatment models and methods of the simulation concerning austenite decomposition kinetics during cooling was carried out to choose the method of forecasting the forming microstructure in a welded joint at EBW. On the basis of the analysis, two types of models were selected for their subsequent adaptation in relation to the thermal cycles of EBW:

- semi-empirical models for constructing diagrams of isothermal and thermo-kinetic decomposition of austenite, developed in [2, 3], and then particularized by the author of [4].
- regression models of transformation of supercooled austenite at continuous cooling, obtained by the authors of [5] with the use of artificial neural networks.

On the basis of these models, one method of forecasting the microstructure of welded joints at EBW is proposed.

The method applies a semi-empirical model for constructing diagrams of austenite isothermal decomposition. This model is based on a general equation for describing austenite transformation curves in a diagram. The equation calculates the time τ required for the formation of ferrite, pearlite and bainite at the temperature T:

$$\tau(X,T) = \frac{F(C)}{F(G)D\Delta T^q}\int_0^X \frac{dX}{X^{2(1-X)/3}(1-X)^{2X/3}}, \quad (1)$$

where $F(G) = \beta 2^{(G-1)/2}$, β – empirical coefficient; G – austenitic grain size (according to ASTM); D – effective diffusion coefficient; ΔT – overcooling; q – indicator, which depends on the effective diffusion mechanism; $F(C,Mn,Si,Cr,Mo)$ – function of the steel composition, expressed in % by weight; integral describes the rate of the reaction of volume fraction transformation and takes into account the sigmoidal effect of phase transformation.

After testing a number of physical models for constructing isothermal and thermo-kinetic diagrams of austenite decomposition, as well as the relations required for their construction, the equations with particularized empirical coefficients were suggested in the studies. The suggested equations, involving

IOP Conf. Series: Materials Science and Engineering **759** (2020) 012020 doi:10.1088/1757-899X/759/1/012020

the determination of the equilibrium temperatures of austenite transformation A_1 and A_3, the determination of the start temperatures of bainite and martensitic transformations, the calculation of the time of ferrite, perlite and bainite formation, more accurately describe isothermal diagrams of austenite decomposition for low- and medium-alloyed steels.

With the use of these equations the austenite decomposition diagram is plotted for a qualitative estimation of the forming structure at EBW, and the calculated cooling curves for the appropriate thermal cycle are marked on it.

The transformations of the semi-empirical model (1) were performed, taking into account the additivity rule, in order to forecast the microstructure quantitative composition of the welded joints and to construct structural diagrams of austenite decomposition for a given thermal cycle. After these transformations the equations determining the volume fraction of the formed ferrite, perlite, and bainite for a given thermal cycle were obtained. In general, these equations are represented as follows:

$$\int_0^X \frac{dX}{X_i^{2(1-X_i)/3}(1-X_i)^{2X_i/3}} = \frac{F(G)}{F(C)} \cdot F(T_i) \cdot \tau_i, \qquad (2)$$

where $F(T_i) \cdot \tau_i = D(T_i) \cdot \Delta T(T_i)^q \cdot \tau_i$, and it is calculated taking into account the numerical data variety including temperature and time variations determined from the thermal welding cycle for a specific section of the welded joint, X_i is a volume fraction of the forming structure at the i moment of time of this cooling curve.

The values of a volume fraction of the formed microstructure X_i are determined in the MathCAD mathematical package with the use of *Given* and *Find* functions, the equation (2) being solved as a system of equations for each i point. The graphs showing the kinetics of the ferrite, perlite, bainite and martensite formation in the cooling process on a given thermal cycle are the result of the salvation of these equations (figure 4). On the basis of the obtained results, a diagram representing the percentage of structural components formation in dependence on the cooling time is constructed (figure 5).

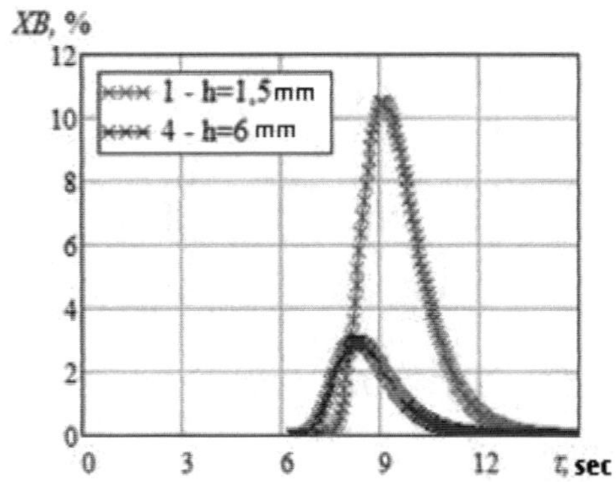

Figure 4. Kinetic diagrams of the formation of bainite depending on the cooling time, calculated from the thermal cycles for the weld metal (points 1 and 4 in Figure 1).

Quantitative estimation was performed with the use of the VideoTest Metal software, which allows painting the phases in accordance with their brightness range and determine the volume fraction of the selected phase. A quantitative estimation of the structural components in the weld metal and the heat-affected zone was carried out to verify the adequacy of the suggested methods for forecasting the microstructural composition. In order to differentiate structural components that are close in morphological structure, such as bainite and martensite, the surface of microsections was alternately treated with two reagents and it was multiple repolished. The analysis was carried out on an optical microscope with the use of polarized light (figure 6).

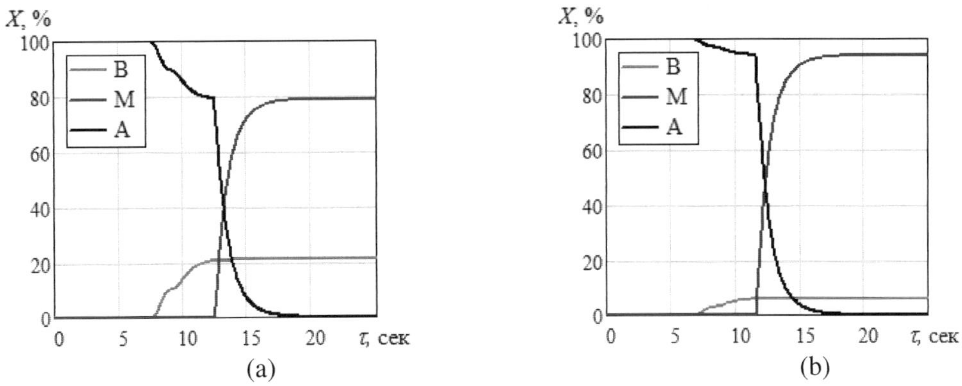

Figure 5. Structural diagrams for weld metal, steel 20Cr3MoNbW, (a) - depth h = 1.5 mm, (point 1 in Figure 1); (b) - depth h = 6 mm, (point 4 in Figure 1); B – bainite, M – martensite, A – austenite.

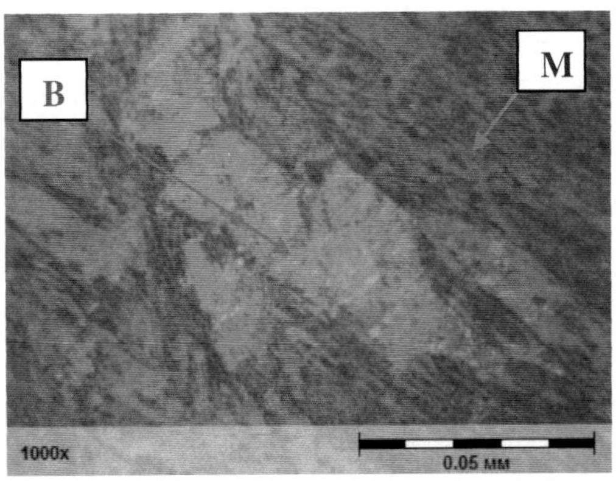

Figure 6. Microstructure of welded joint metal (Figure 1), 20Cr3MoNbW steel, weld upper part, B - bainite, M – martensite.

Table 2 presents the experimentally determined and calculated structural composition of the welded joint obtained on the basis of the physical model of the suggested method. The method allows making a quantitative prognosis of the microstructure of the weld metal and the heat-affected zone at EBW of low- and medium-alloyed steels.

Table 2. Calculated and experimental structural compositions of the welded joint, the steel 20Cr3MoNbW, Figure 1.

Analyzed weld sections					
Weld metal,		Overheating zone,		Zone of complete recrystallization,	
h= 1,5 mm	*h*= 6 mm	*h*= 1,5 mm	h= 6 mm	*h*= 1,5 mm	*h*= 6 mm
21 % B, 79 % M	5,9 % B, 94,1% M	14,4 % B, 85,6 % M	11,6 % B, 88,4 % M	11 % B, 89% M	9,7 % B, 90,3 % M

Weld metal		Heat-affected zone	
h = 1,5 mm	*h* = 6 mm	*h* = 1,5 mm	*h* = 6 mm
21,1 % B 78,9 % M	5,5 % B 94,5% M	15,9 % B 84,1 % M	11,8 % B 88,2 %M

4. Conclusion

Thus, the method of forecasting the microstructure formed at EBW of medium-alloyed steels was suggested taking into account special characteristics of thermal cycles variations along the width and the depth of a welded joint. The suggested method is based on the construction of the structural diagrams describing the austenite transformation kinetics during the continuous cooling in the dependence on the thermal cycle of EBW.

References

[1] Olshanskaya T V, Fedoseeva E M, Koleva E G 2017 Creation of thermal models at electron beam welding by method of functions of green *Bulletin of PNRPU. Mechanical engineering, materials science*, **19 (3)** 49-74

[2] Karhin V A 2013 *Welding Heat Processes* (Saint-Petersburg: Polytechnic University Publishing House)

[3] Kirkaldy J S 1991 Diffusion-controlled phase transformations in steels. Theory and application *Scandinavian Journal of Metallurgy* **20** 50

[4] Li M, Niebuhr D, Meekisho L, Atteridge D 1998 A computational model for the prediction of steel hardenability *Metallurgical and Materials Transactions* **29** 661-671

[5] Dobrzański L A, Malara S, Trzaska J 2007 Application of the neural network for Mg-Al-Zn mechanical properties modeling *Journal of Achievements in Materials and Manufacturing Engineering* **1-2** 351-354

Numerical model of a hollow cathode arc discharge formation in vacuum

G L Permyakov[1,5], **D N Trushnikov**[1], **M K Mindibaev**[1], **E G Koleva**[2,3,4]

[1] Perm National Research Polytechnic University, 29 Komsomolsky ave., Perm, 614990, Russian Federation
[2] Institute of electronics, Bulgarian Academy of Sciences, 72, Tzarigradskoye Highway, Sofia 1784, Bulgaria
[3] Technological Center on Electron Beam and Plasma Technologies and Techniques – TC EPTT Ltd., Bulgaria
[4] University of Chemical Technology and Metallurgy – Sofia, 8, Kliment Ohridski blvd., Sofia 1756, Bulgaria
[5] E-mail: gleb.permyakov@yandex.ru

Abstract. Industrial application of hollow cathode arc (HCA) discharge necessitates understanding the processes of interaction between the plasma and the cathode and the anode, as well as related processes. The first step is to study the formation process of HCA discharge in vacuum with the micro-flow of plasma-forming gas through the hollow cathode. In this work presents a two-dimensional model describing the related processes of transfer of charged particles and the movement of plasma-forming gas flows. Electron density and mean electron energy are calculated by solving the drift-diffusion equations. The mass transfer equation for a multicomponent mixture is used to describe the mass transfer of heavy plasma particles. To calculate the electric field strength the Poisson equation is used. The emission of secondary electrons from the inner surface of the cathode is taken into account. The boundary conditions take into account the loss of charge as a result of chaotic motion and its occurrence due to thermal emission effects. The gas flow is determined by collisions and diffuse re-reflection from all surfaces assumed in accordance with Knudsen's law. Calculations of plasma formation and the movement of plasma-forming gas flows were performed using the simulation package COMSOL Multiphysics.

1. Introduction

One of the most common sources of energy for welding, surfacing and related technology is electric arc discharge. The combination of the efficiency of vacuum protection and the properties of the electric arc as an energy source for creating arc welding processes in vacuum on their basis seems very promising. This technical problem served as the basis for the development of arc welding processes in vacuum. As a result a scheme was chosen for producing an arc discharge with a non-consumable thermionic hollow cathode in vacuum, the main advantage of which is the possibility of obtaining, under certain conditions, a low value of specific erosion.

Industrial application of hollow cathode arc (HCA) discharge necessitates understanding the processes of interaction between the plasma and the cathode and the anode, as well as related processes. The theory acquires a particularly important role in the development of new arc processes under specific conditions different from traditional ones, since in these cases the experiment alone often does not make it possible to carry out the required process. The first step is to study the formation process of HCA discharge in vacuum with the micro-flow of plasma-forming gas through the hollow cathode. When studying arc discharges in vacuum, the fundamental work of physicists involved in the theory of gas discharges was used [1-4].

2. Mathematical model

In this work presents a 2D model to investigate the HCA discharge in vacuum when argon used as the plasma-forming gas. The calculations of plasma formation and the movement of plasma-forming gas flows were performed using the simulation package COMSOL Multiphysics.

Figure 1 shows a scheme of the model (calculations are performed in cylindrical coordinates in an axisymmetric formulation). Hollow cathode parameters: inner radius r_c=1.5 mm, cathode thickness δ_c =1 mm, length l_c=10 mm.

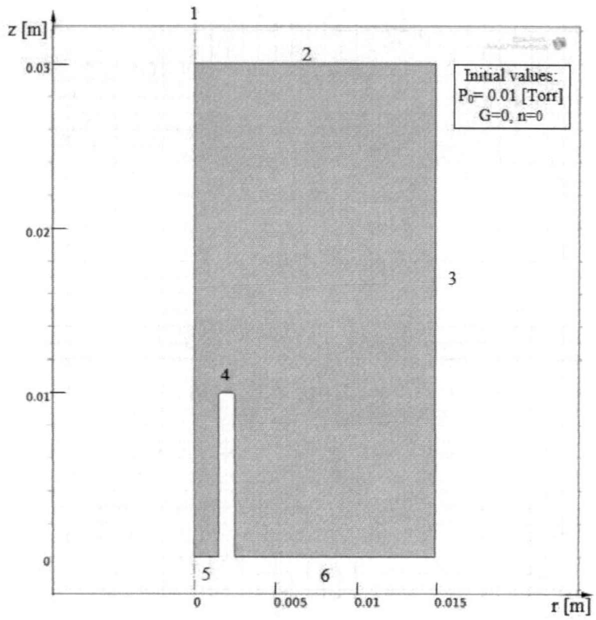

Figure 1. Schematic of the HCA discharge model: 1 – axis of symmetry, 2 – anode (U=0), 3 and 6 – gas outlet (S=100 [l/s]), 4 – cathode (U=U_0), 5 – gas inlet (Q_m).

3. Governing equations

Electron density and mean electron energy are calculated by solving the drift-diffusion equations [1, 5, 6]:

$$\frac{\partial n_e}{\partial t} + \nabla \cdot \mathbf{G}_e + (\boldsymbol{u}_g \cdot \nabla) n_e = \boldsymbol{R}_e \tag{1}$$

$$\frac{\partial n_\varepsilon}{\partial t} + \nabla \cdot \mathbf{G}_\varepsilon + \mathbf{G}_e \cdot \boldsymbol{E} + (\boldsymbol{u}_g \cdot \nabla) n_\varepsilon = \boldsymbol{R}_\varepsilon \tag{2}$$

where n_e and n_ε are the electron density and electron energy density, respectively. $\mathbf{G}_e$ and $\mathbf{G}_\varepsilon$ are the electron flux and electron energy flux. The collision source term R_e represents the production or loss of electrons, and the electron impact reactions are summarized in Table 1 [7-9]. In these equations, R_ε represents the electron energy change in those reactions, $\boldsymbol{u}_g$ is the flow velocity of neutral gas, and $\boldsymbol{E}$ is the electric field intensity.

Table 1. Plasma chemistry reactions.

№	Formula	Data of reaction	Type of interaction
1	e + Ar → Ar + e	cross-section	elastic
2	e + Ar → Ar* + e	cross-section	excitation
3	e + Ar* → Ar + e	cross-section	excitation
4	e + Ar* → Ar⁺ + 2e	cross-section	ionization
5	e + Ar → Ar⁺ + 2e	cross-section	ionization

The mass transfer equation for a multicomponent mixture is used to describe the mass transfer of heavy plasma particles [10–12]:

$$\rho\frac{\partial\omega_k}{\partial t} + \rho(\boldsymbol{u}_g\cdot\nabla)\omega_k + \nabla\cdot\mathbf{G}_k = \boldsymbol{R}_k \quad (3)$$

where ρ is the mass density of the working gas, ω_k is the mass fraction of the kth species, $\boldsymbol{R}_k$ is a collision term of the heavy particles, derived from the plasma chemical kinetic reactions in Table 1, $\mathbf{G}_k$ is the flux of the kth species.

To calculate the electric field intensity, the Poisson's equation is used:

$$-\varepsilon_0\nabla\cdot(\nabla U) = \varepsilon_0\nabla\cdot\boldsymbol{E} = \rho \quad (4)$$

where U is the electrostatic potential of plasma, ρ is the volume density of the charge, and ε_0 is the dielectric constant of vacuum.

Discharge in plasma is accompanied by the emission of secondary electrons from the surface of cathode. The boundary conditions take into account the loss of charge as a result of chaotic motion and its occurrence due to thermal emission effects [11]:

$$-\boldsymbol{n}\cdot\mathbf{G}_e = \left(\frac{1}{2v_{e,t}n_e}\right) + n_e\mu_e\boldsymbol{E}\cdot\boldsymbol{n} - \boldsymbol{n}\cdot\mathbf{G}_t \quad (5)$$

and for the electrons energy flux:

$$-\boldsymbol{n}\cdot\mathrm{G} = \left(\frac{1}{2v_{e.t}n_\varepsilon}\right) + n_\varepsilon\mu_\varepsilon\boldsymbol{E}\cdot\boldsymbol{n} \quad (6)$$

where $v_{e.t}$ is the thermal velocity of electrons, $\mathbf{G}_t = \boldsymbol{J}_t/e$ is the thermal electron emission flux and $\boldsymbol{n}$ is the surface normal, μ_e and μ_ε are electron mobility and electron energy diffusion coefficients.

Boundary condition for electron flux, electron energy flux and heavy particle flux on edges 2, 3, 4, 5, 6 detailed in [13].

The gas flow is determined by collisions and diffuse re-reflection from all surfaces assumed in accordance with Knudsen's law using Free Molecular Flow module of COMSOL Multiphysics.

4. Results

Figure 2 shows the calculated electron density distribution during HCA discharge in vacuum with the micro-flow of plasma-forming gas (Argon) through the hollow cathode cavity. Preliminary calculations have shown that the maximum of electron density is located near the hollow cathode tip.

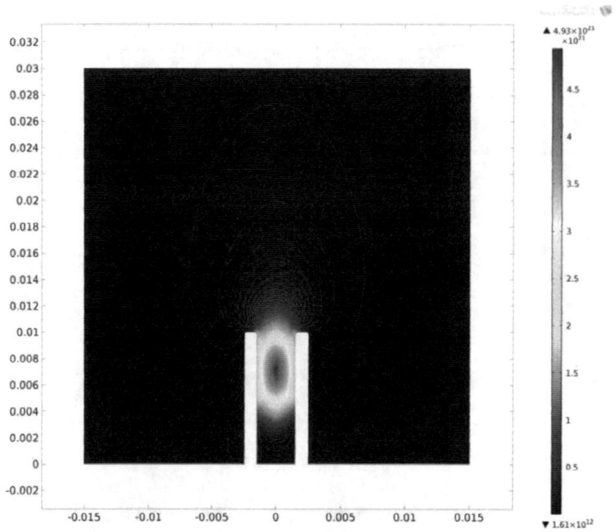

Figure 2. Electron density distribution during HCA discharge (discharge current Id = 96A, electric potential U0 = 30 V, gas flow Qm = 3 mg/s).

5. Conclusion

This work presents a model of the HCA discharge describing the related processes of plasma formation, transfer of charged particles and the distribution of plasma-forming gas.

Qualitatively, the results do not contradict the known theoretical calculations and experiments. At further research should to take into account heat and mass transfer in the process of forming an HCA discharge and study this question in detail also for use of the developed model, it is necessary to calibrate it and perform verification based on experimental data.

6. Acknowledgments

The reported study was supported by the Government of Perm Krai research project Nr. S-26/795 of December 21, 2017, by the grant from the Russian Foundation for Basic Research RFBR Nr. 18-08-01016 A, as well as by Ministry of Science and Higher Education at the base part of the state assignment Nr. 9.9697.2017/8.9.

References

[1] Raizer Yu. P., Allen J. E. 1991 *Gas Discharge Physics* (Berlin: Springer-Verlag)
[2] Lafferty J M, Cobine J D 1980 *Vacuum Arcs: Theory and Applications* (New York : Wiley)
[3] Granovsky V L 1971 *Electric discharge in gases* (Moscow: Nauka) [In Russian]
[4] Kesasv I G 1968 *Cathode processes of an electric* arc (Moscow: Nauka) [In Russian]
[5] Richards A D, Tompson B E and Sawin H H 1987 Continuum modeling of argon radio frequency glow discharges *Applied Physics Letters* **50 (9)** 492–494
[6] Lieberman M A. Lichtenberg A J 2005 *Principles of Plasma Discharges and Materials Processing* (New Jersey: John Wiley & Sons)
[7] Raju G G 2006 *Gaseous Electronics: Theory and Practice* (Florida: CRC Press)
[8] Lee C, Lieberman M A 1995 Global model of Ar, O2, Cl2, and Ar/O2 high-density plasma discharges *Journal of Vacuum Science & Technology A* **13** 368–380
[9] Denysenko I, Stefanovi I, Sikimi B, Winter J, Azarenkov N A and Sadeghi N 2011 A global model for the afterglow of pure argon and of argon with negatively charged dust particles *Journal of Physics D: Applied Physics* **44 (20)** 205204

[10] Bird R B, Stewart W E and Lightfoot E N 2002 *Transport Phenomena* (New Jersey: John Wiley and Sons)
[11] Hagelaar G J M 2000 *Modeling of microdischarges for display technology* (Eindhoven: Technische Universiteit Eindhoven)
[12] Brokaw R S 1969 Predicting transport properties of dilute gases *Industrial & Engineering Chemistry Process Design and Development* **8 (2)** 240–253
[13] Cong S, Wu R, Mu L, Sun J and Wang D 2019 A 2D model for low-pressure hollow cathode arc discharges in argon *Journal of Physics D: Applied Physics* **52 (4)** 045205

Computer simulation of metals ionization process during electron-beam welding

R V Rodyakina, A V Shcherbakov and D A Gaponova

National Research University "Moscow Power Engineering Institute", 14, Krasnokazarmennaya str., Moscow, 111250, Russia

E-mail: reginarodyakina@mail.ru

Abstract. One of the most actual problems during electron beam welding of metal materials is the determination of the laws of primary electrons interaction with a gas (vapor) flow, taking into account existing and generated (by charged particles) electric fields. The paper is devoted to the development of computer simulation method for the process of electron beam passage through a layer of evaporated metal. The analysis of simulation results showed a significant effect of ionization process on electric field in a gas-vapor channel. It was established that at the initial stage of the process, a region with a positive potential is formed in the lower part of gas-vapor channel, and in its upper part, on the contrary, potential decreases.

1. Introduction

Currently, many areas of machine-building production are intimately associated with making such products that are subject to increased requirements for strength, ductility, corrosion resistance, etc. One of the promising areas that can successfully overcome different technological difficulties arising during the process of such products making is electron beam processing. Electron beam heats metal (during electron deceleration), causes its melting and evaporation.

One of the most actual tasks is to determine the laws of interaction of primary electrons with a gas (vapor) flow, taking into account existing electric fields and generated by charged particles. Therefore, this paper is devoted to the development of computer simulation method for the process of electron beam passage through a layer of evaporated metal.

2. Method of computer simulation

As the initial data for modeling, it is necessary to set medium density or concentration of particles η. It is known [1] that density of particles above the surface of penetration channel is about 10^{17} per 1 cm^3, and above the surface of molten metal (at the bottom of the channel), it approaches to the density of particles in the solid state, i.e. it is about 10^{24} in 1 cm^3. The computer simulation process was performed for aluminum.

In addition to ionization, the process of electron passage through a vapour layer will be accompanied by energy losses for other types of interactions, for example, excitation of phonons, plasmons, deceleration radiation generation and recombination processes. In addition, it is possible to change the direction of primary electrons motion due to elastic interactions [2]. The purpose of this study is to analyze ionization processes, potential distribution and ionization losses.

In the simulation process, we shall consider inelastic interaction cross-section — ionization cross-section of gas (vapour) molecules by electron impact. As is known, ionization cross-section of an atom or molecule by electron impact is a function of electron energy [3]. Typically, ionization cross-section is about 10^{-16}–10^{-17} cm^2. The dependence of ionization cross-section of aluminum atom (σ_{Al}) on electron energy used in the proposed model is shown in Figure 1. The model diagram is shown in Figure 2. Target 1 (product) of aluminum has a cylindrical shape, its lateral and lower faces are grounded (their potential is taken equal to zero). The gas-vapor channel in the model is approximated by normal distribution function (item 2 in Figure 2). The distribution of vapors density spreading in vacuum 3 in the channel and beyond it is specified. Potential on the channel wall and on the upper surface of cylinder is determined by field potential created by electron beam 4. Electron flux density obeys the normal distribution function. Ionization processes occur in the vapor-gas channel, and charged particles move under the action of Lorentz forces in the field, created by the particles themselves. This field does the work and changes trajectories of the particles.

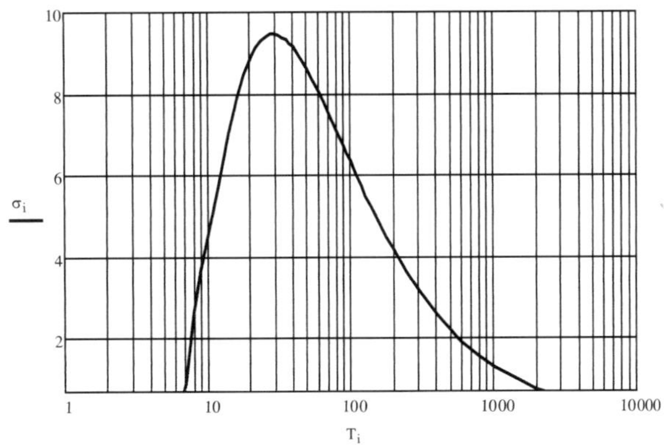

Figure 1. Dependence of ionization cross-section for aluminum atom on electron energy: T_i – electron energy values, eV; σ_i – values of ionization cross-section, × 10^{16}, cm^2 [7]

The calculation of electric field is carried out in an axially symmetric formulation. For this case, the equation can be used

$$\frac{\partial^2 U}{\partial r^2}+\frac{1}{r}\frac{\partial U}{\partial r}+\frac{\partial^2 U}{\partial z^2}=-\frac{\rho(r,z)}{\varepsilon_0}, \tag{1}$$

where U is electric potential (V); r and z are radial and axial coordinates, respectively (m), ε_0 is permittivity of vacuum (F·m^{-1}); $\rho(r,z)$ is volume density of spatial charge created by charged particles beam in the zone of field action (C·m^{-3}). Its equation can be written as

$$\rho=\frac{dQ}{dV}, \tag{2}$$

where dQ is charge part, distributed over the elementary volume dV. A cylindrical grid is superimposed on two-dimensional computational region, cells of which are numbered along r-axis as $i = 0,1,2 \ldots i_{\max}$, and along z-axis as $j = 0,1,2 \ldots j_{\max}$ (Figure 2). The sizes of cylindrical grid cells are Δr and Δz, respectively. In accordance with selected grid scheme, calculation values $U_{i,j}$ are performed in consecutive order for all nodes of the grid and are stored in the data array $U(i,j)$. A finite-difference approximation of computational region was used, for which equation (1) can be written as

$$\frac{U_{i+1,j}-2U_{i,j}+U_{i-1,j}}{\Delta r^2}+\frac{1}{i\Delta r}\frac{U_{i+1,j}-U_{i,j}}{\Delta r}+\frac{U_{i,j+1}-2U_{i,j}+U_{i,j-1}}{\Delta z^2}=-\frac{\rho_{i,j}}{\varepsilon_0}. \tag{3}$$

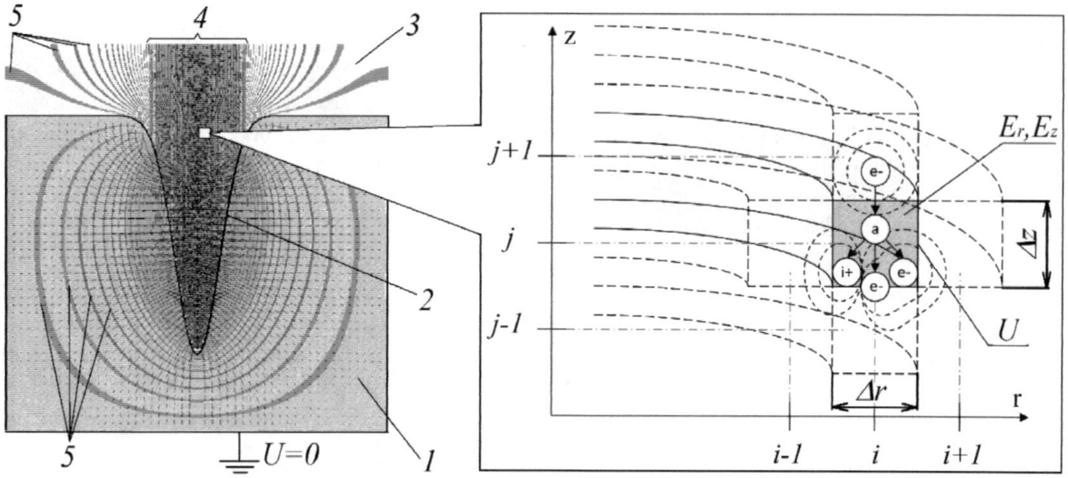

Figure 2. Diagram of two-dimensional model of the process of metal vapor ionization in a gas-vapor channel: 1 - grounded product; 2 - gas-vapor channel; 3 - vacuum, metal vapor; 4 - electron beam; 5 - equipotentials; elements of the computational grid and diagram of electron impact ionization process are shown on the right

This equation was solved by the method of Euler. Then, using grid function $U(i,j)$ obtained, vector field of electric field strength E (V/m) was calculated by numerical differentiation

$$\vec{E} = -\mathrm{grad}U\,, \tag{4}$$

as well as projections of this vector on the coordinate axes E_r and E_z for each node of the computational grid. The acceleration components, acquired by a particle when moving in an electrostatic axially symmetric field due to the Coulomb force, are determined from the equatuion

$$\begin{cases} a_r = \dfrac{qE_r}{m} \\ a_z = \dfrac{qE_z}{m} \end{cases}, \tag{5}$$

where a_r and a_z are the radial and axial components of acceleration (m·s^{-2}); q is particle charge (C); m is particle mass (kg). The task of particle trajectory calculating is solved with a fixed time integration step (which was taken equal to 10^{-12} s). Equations for the projections of speeds increments Δv_r, Δv_z and coordinates increments ΔS_r, ΔS_z of the particle in the time interval (τ_1, τ_2) are written as follows:

$$\Delta v_r = \frac{q}{m}\int_{\tau_1}^{\tau_2} E_r dt\,,\quad \Delta S_r = \int_{\tau_1}^{\tau_2} v_r dt\,, \tag{6}$$

$$\Delta v_z = \frac{q}{m}\int_{\tau_1}^{\tau_2} E_z dt\,,\quad \Delta S_z = \int_{\tau_1}^{\tau_2} v_z dt\,. \tag{7}$$

The distribution of current density in the product was calculated on the basis of Ohm's law in differential form

$$\vec{j} = \gamma\vec{E}\,, \tag{8}$$

where $\vec{j}$ is current density vector (A·m^{-2}), γ is specific conductivity of product's material (aluminum), Ohm^{-1}·m^{-1}.

The most difficult stage of the problem to solve is simulation of ionization process, which is accompanied by energy exchange between particles and appearance of a free electron. In accordance with definition of the term "interaction cross-section" [4], mean free path of an electron λ_i between ionization acts is defined as

$$\lambda_i = \frac{1}{\eta \sigma_i}, \quad (9)$$

where η is concentration of aluminum atoms (m^{-3}). Concentration of atoms is related to vapor pressure by ideal gas law

$$\eta = \frac{P}{kT}, \quad (10)$$

where T is vapor temperature, k is Boltzmann constant. At the current stage of research, a linear approximation of pressure distribution function over the channel depth $P(z)$ was used, which is agree with published works [5].

To take into account random nature of vapour particles distribution, the Monte Carlo method [6] was used. In this case, path s_i, which electron passes between ionization acts, will be determined by the equation

$$s_i = -\lambda_i \ln(1 - R_1), \quad (11)$$

where R_1 is a dimensionless random value, uniformly distributed in the range from 0 to 1. As a result of ionization, a pair of such particles, as positive ion and slow electron, appears; their trajectories will also need to be calculated in each time interval.

The initial energy of secondary electron W_s (eV), which appeared as a result of ionization, is proposed to determine by the equation

$$W_s = \omega \cdot tg \left\{ R_2 arctg \left(\frac{W_p - J}{2\omega} \right) \right\}, \quad (12)$$

where W_p is energy of bombarding (primary) electron (eV); J is ionization potential; ω is tabular value, obtained by comparing theoretical and experimental data (for aluminum this value is equal 2.7 eV); R_2 is a dimensionless random value, uniformly distributed in the range from 0 to 1.

In the case considered above, primary electron energy loss due to ionization ΔW_{pi} (eV), was calculated by the equation

$$\Delta W_{pi} = J + W_s. \quad (13)$$

Path that an electron travels between elastic interaction acts s_e is determined by the equation

$$s_e = -\lambda_e \ln(1 - R_3), \quad (14)$$

where R_3 is a dimensionless random value, uniformly distributed in the range from 0 to 1.

In order to study ionization processes, calculations in accordance with expressions (1)–(14) were implemented in the Elion program module in C++.

3. Results of simulation and their discussion

At the current stage of research, the most interesting was information on the distribution of electric field potential, created by particles in the penetration channel and near it, as well as spatial distribution of particle ionization density acts. Figure 3 shows the results of computational experiments, realized on a personal computer with an IntelCorei7-6700 HQ processor.

Figure 3a shows the case corresponding to the absence of aluminum vapor concentration (it is equal to zero) in the region above penetration channel. This situation corresponds to the initial stage of electron beam impact on the target (in the real case, penetration channel is not formed yet, and sample surface is flat). The beam current is 18 mA. It is seen that a region with the smallest electric potential is in the beam center. In the approach to the bottom of the channel, potential increases, approaching almost to zero, since potential difference between the channel bottom surface and the lowest (grounded) plane of the product is negligible for this case. The minimum potential value under electron beam steady action on the product without taking into account beam interaction with metal vapors is equal to –1.8 V. In the case when vapor pressure distribution of aluminum along *Z*-axis is specified, the picture changes radically: potential distribution will be completely different after a time on the order of 0.05 ns. Figure 3b shows such a case. Vapor pressure of aluminum in the lowest part of the channel is 30 Pa, and at the point $z = 0$ (at the distance of 2 mm above the product surface) it is equal to 0.1 Pa. The dependence $P(z)$ at this stage of the research was assumed to be linear.

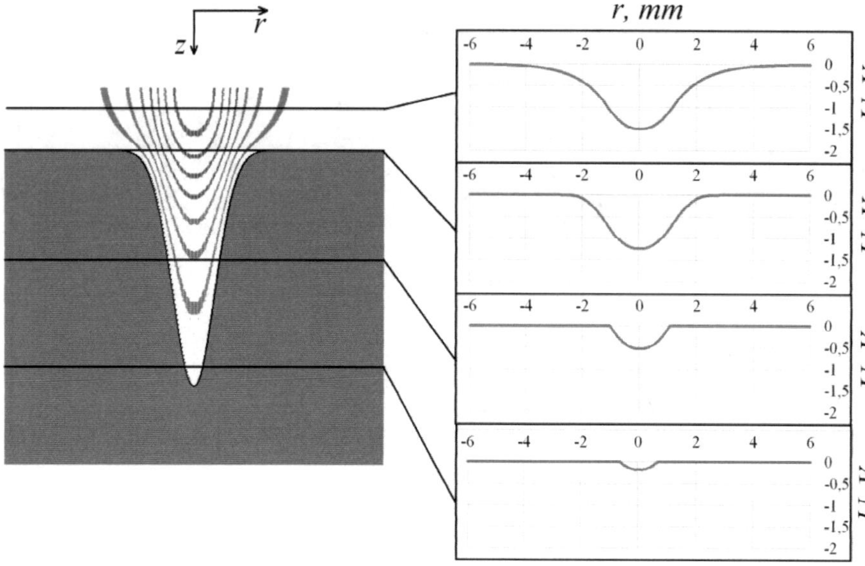

a)

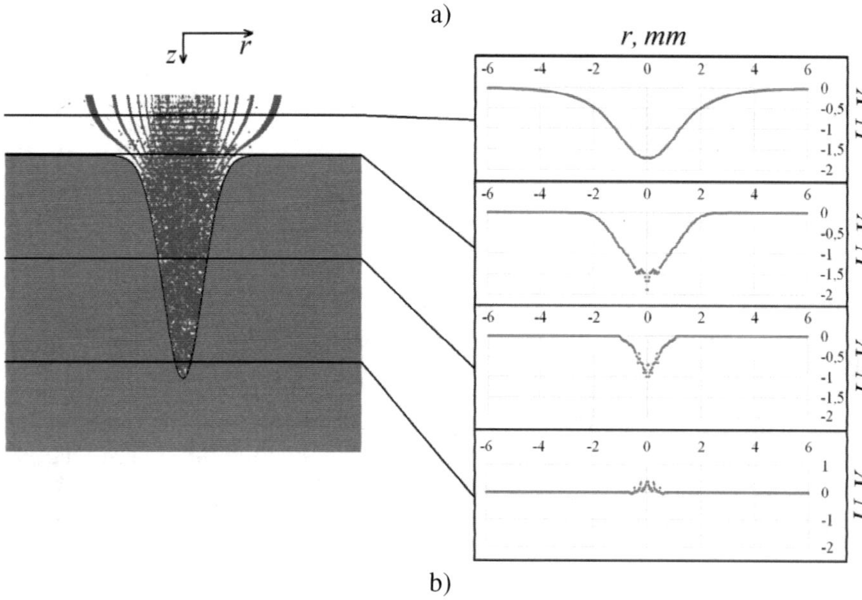

b)

Figure 3. Electric field distribution: a) in the channel in the absence of metal vapor, b) taking into account linear distribution of vapor pressure along coordinate *z*.

Ionization of neutral aluminum atoms leads to the appearance of positively charged ions and slow electrons. Due to the fact that electron beam has a current density distribution over its radius, most of ions appear precisely in the center of the beam, and at the initial stage of ionization process, shown in Figure 3,b, slow electrons will also appear near the center of the beam. Under the action of positive ions field, localized mainly in the center of the beam, peripheral slow electrons will drift in their direction, creating a sharp minimum of potential in the center.

In the lower part of the channel, on the contrary, a region with a positive space charge will be formed, as evidenced by the fact that potential here will assume a positive value. It can be predicted that, under the action of primary and slow secondary electrons, a region with a potential gradient sufficient to accelerate ions in the direction of electron gun will be formed, and forming plasma will correspond in its characteristics (density and temperature) to a glow-discharge plasma. However, processes of ion acceleration simulation using the developed program, run on a computer with the above-cited characteristics, may take considerable time (up to several days). This is due to a significant difference in the masses of ions and electrons. The most obvious solution of this problem is transition to parallel computing technology.

4. Conclusion

1. It is shown that, using proposed complex physical and mathematical model, it is possible to study plasma formation process during the influence of electron beam on metals.
2. Analysis of the first simulation results has been carried out. There was shown a significant effect of ionization process on electric field distribution in gas-vapor channel. It was established that at the initial stage of the process, a region with a positive potential is formed in the lower part of gas-vapor channel, and in its upper part, on the contrary, potential decreases.
3. It is shown that for effective application of the developed model and studying of the drift process of metal ions, having a significant mass in comparison with electrons, using of parallel computing technology is necessary.

5. Acknowledgments

The research was carried out in National Research University “Moscow Power Engineering Institute” with the financial support of the Russian Science Foundation (project No. 18-19-00652).

References

[1] Novokreshchenov V V, Rodyakina R V and Karimbekov M A 2017 *Solid State Phenomena* **265** 237–45
[2] Nazarenko O K, Kaydalov A A and Akopyants K S 1973 *Automatic welding* **8** 75—6.
[3] Bondarenko G G, Kristia V I and Fisher M R 2005 *Interaction of Ions with the Surface: The XVII Int Conf (Moscow)* vol 2 331–4.
[4] Nazarenko O K, Kaydalov A A and Kovbasenko S N 1987 *Electron-beam Welding* (Kiev: Naukova Dumka) p 256
[5] Rai R. 2008 *Modeling of Heat Transfer and Fluid Flow in Keyhole Mode Welding (A Diss. in Materials Science and Engineering)* (USA: The Pennsylvania State University) p 206
[6] Kvlividze V A and Krasilnikov S S 1985 *Introduction to the physics of atomic collisions* (Moscow: Publishing House of Moscow State University) p 224
[7] Kim Y K and Stone P M 2001 *Phys. Rev.* A **64** 052707

Threshold conditions for thermocapillary transition to deep penetration mode in selective laser melting of metal powder bed

R D Seidgazov [1], **F Kh Mirzade** [1]

[1] ILIT RAS - Branch of the FSRC "Crystallography and Photonics" RAS, Svyatoozerskaya, 1, 140700, Shatura, Moscow Region, Russia

E-mail: seidgazov@mail.ru

Abstract. The keyhole mode of selective laser melting (SLM) of metal powder bed in additive technology is characterized by an intensive hydrodynamic process in a thin molten layer. Such a mode is widely used also in laser and electron beam welding indicating the similarity of hydrodynamic processes in these technologies despite a significant difference of operating parameters. The threshold conditions of thermocapillary keyhole mode transition for various metals (Cu, Fe, Ti) in a wide range of beam parameters used for selective laser melting of metal powder layer and laser welding are investigated. The condition of threshold for thermocapillary transition into keyhole mode by sticking of viscous layer to the solid boundary is formulated. The fulfillment of this condition is confirmed by the convergence of estimated values of the viscous layer thickness and the molten layer depth during the transition to a keyhole mode. Analytical estimates of keyhole threshold and comparisons with experimental values of beam power and spot size corresponding to the transition in keyhole mode for SLM processes and laser welding are presented. The correlation of these values confirms the thermocapillary mechanism of cavity formation and the similarity of hydrodynamic processes in laser welding and SLM processing in keyhole mode at wide range of operating parameters.

1. Introduction

The technology of additive manufacturing with selective laser melting (SLM) of metal powders is used for 3D printing of metal products with CAD-modeling of the process. SLM technology allows creating precise metal parts of components and assemblies by power laser beam successfully replacing in some cases traditional methods of manufacturing.

A 3D part produced by SLM technology is a set of layers (thickness of the order of 20 - 100 μm) fused from particles of a metal powder and deposited sequentially on top of one another. Sintering is carried out under the local heating by a laser beam moving over the selected zone, melts the powder particles. The molten powder material then solidifies in the form of thin strips. After the formation of the first zone is completed, a new layer of powder is applied to the entire surface of the sample and the process is repeated. Its undoubted advantage is the ability to use a fairly wide range of source powders (from plastic to metals and alloys) for the manufacture of products.

The additive production of products by SLM is determined by a number of operating parameters, such as a beam power, rate and strategy of scanning, spot size, physical and dispersion properties of the powder, etc. It should be noted the interconnection of factors affecting the process and the presence of many interacting processes: absorption and scattering of laser radiation energy by the substrate substance and powder particles, heat distribution and convection, evolution of the free surface of the melt strip due to capillary forces, evaporation, shrinkage, crystallization, formation microstructures of the synthesized object. The formation of high-quality layers is associated with the search for the optimal range of technological parameters of the process, which is usually quite narrow.

Layers melting can be carried out in deep penetration mode, when the depth of melt L exceeds the width D ($L>D$). This mode is widely used in welding by laser or electron beam. The difference lies in a significant difference in the characteristic scales and parameters values of processes in SLM and welding by laser or electron beam. The SLM technology uses a beam spot sizes of ~ 50 microns in size. And the characteristic diameter of beam spot used in keyhole welding can be ~ 200 ÷ 500 microns and even more. The same difference is the depth of penetration. The penetration depth usually varies within $L \sim 1 \div 100$ mm in laser/electron beam welding, but is comparable to the powder layer thickness of ~ 0.1 mm in SLM.

The quality of products produced by SLM technology is largely related to hydrodynamic phenomena in laser melting, which occur particularly intensively in deep penetration (keyhole) mode. This determines the interest in such phenomena in SLM [1-7].

The transition from the low characteristic melting depth ($L < D$) conduction mode to the deep penetration mode ($L > D$) is accompanied by a sharp increase in the crater depth and the beam penetration deep into the material after reaching a threshold power density values.

It is often assumed that the reason for threshold growth of penetration depth is a sharp increase in vapor pressure under influence of high-power focused radiation, which is supplemented by radiation capture in cavity with an increase of effective absorption due to multiple reflections inside cavity [8,9,10]. This assumption (let's call it the evaporative-hydrodynamic hypothesis) is widely used for about half a century (beginning from first theoretical considerations [11-13]) in the modeling of the deep penetration process in relation to beam technologies of metal melting, although it does not have up to now the necessary experimental evidence. Therefore, the evaporative mechanism of deep penetration phenomenon under the action of high vapor pressure reflects the established tradition rather than a scientifically substantiated fact. As for a wide range of specialists the evaporative-hydrodynamic mechanism of the melt flow which cause a cavity formation is perceived undeniable, reliably established and even obvious fact. But this mechanism of cavity formation allows creating numerical models of deep penetration under the influence of a laser or electron beam, which with acceptable accuracy reproduce only the thermal processes and the shape of the melted zone that is sufficient for some applications.

However, experts note the difficulties in calculation of hydrodynamic processes associated with an adequateness of understanding the real hydrodynamic mechanisms [14,15]. Such a difficulties appear when detailed reproduction in calculations of deep penetration phenomenon, in prediction of defects of hydrodynamic nature (pores, root defects, hamping, undercuts, perforation, splashing, etc.). The accuracy of predictions based on the evaporative-hydrodynamic hypothesis is of unacceptably low accuracy, which does not allow solving practical problems. The evaporation rate values obtained by Langmuir equation are by order of magnitude higher then measured results [16]. There are other critical discrepancies between this hypothesis and empirical data [17,18], which indicate negligible evaporation inside a keyhole. It should also be noted that representations of evaporative-hydrodynamic hypothesis did not help to obtain exact values of the threshold radiation parameters for deep penetration mode transition.

The reason for these problems sometimes looks for imperfection of existing knowledge about evaporation process [16]. Attempts are also being made to find a solution by expanding the physical phenomena under consideration. From the other side there are a strong doubts about the validity of the evaporative-hydrodynamic hypothesis application in SLM or keyhole welding, since it has no direct evidence by experimental methods [19,20], including evidence of a causal relationship between the increase in evaporation and the increase in the penetration depth.

Therefore, it is important to pay attention to experimental and theoretical data indicating a deep cavity formation under the action of thermocapillary forces at high surface tension gradients in the melting zone [18,19,21,22]. In this case, the transition into deep penetration mode is associated with the restructuring of the thermocapillary flow when the radiation threshold values are reached.

Knowledge of the threshold parameters for deep penetration mode transition is necessary for confident control of laser melting process in SLM and welding technologies. A detailed analysis of the threshold conditions of deep penetration based on the thermocapillary mechanism is presented in this paper. The calculated and experimental threshold radiation parameters for deep penetration transition are compared. The results are considered for a number of metals (Cu, Fe, Ti) in a wide range of threshold parameters, which covers SLM, LW, EBW technologies.

2. Velocity of thermocapillary flow and the thickness of viscous layer during melting in deep penetration mode

Transition from conduction melting mode ($L < D$) to deep penetration mode ($L > D$) associates with a molten layer evolution which is differ at prethreshold conditions of laser melting and above the threshold parameters.

Metal surface temperature increases after beam switching on and reaches the melting point T_M during time t_M . Thin melted layer appeared in irradiation zone and rapidly increases in size. A surface tension gradient $\partial\sigma/\partial T$ and tangential thermocapillary force $\partial\sigma/\partial T\times\partial T/\partial r$ that is directed to reducing the surface tension are formed on thin molten layer surface due to the temperature dependence of surface tension $\sigma(T)$. The surface tension is maximal at the melting point T_M and tends to zero under critical conditions ($\partial\sigma/\partial T < 0$). Therefore, thermocapillary force is directed from the center to the beam spot periphery. Some exception to this rule is possible only in a narrow range (within 200-300 degrees) above T_M, where the influence of impurities and contaminants can affect. In this small range $\partial\sigma/\partial T$ can have a positive sign ($\partial\sigma/\partial T > 0$), and the direction of thermocapillary force can be directed from spot periphery to center. However, keyhole effect is characterized by melt temperatures close to the boiling point T_B , at which there is no influence of contaminants and impurities on the melt hydrodynamics.

The flow appeared is determined by the balance between thermocapillary stress on melt surface and counteracting viscous friction of lower layers

$$\eta\partial V_X/\partial z = |\partial\sigma/\partial T|\partial T/\partial r \tag{1}$$

η - dynamic viscosity, V_X - the tangential velocity of thermocapillary melt flow, r - tangential coordinate z - normal coordinate.

When melting start the liquid layer thickness is always smaller than focus spot diameter and the convective transfer can be neglected. Therefore, the linear Navier-Stokes equations can be used

$$\partial\mathbf{V}/\partial t = -1/\rho\ grad\, p + \nu\Delta\mathbf{V}$$

To exclude pressure, let us apply the rot operation to both sides of this equation. Taking into account $rot\ grad\ p = 0$, receive

$$\partial rot\ \mathbf{V}/\partial t = \nu\Delta rot\mathbf{V}$$

The Navier-Stokes equation in this form is similar to the heat equation. So a characteristic distance of viscous forces propagation can be defined as $2(\nu t)^{1/2}$. If the thickness of the viscous layer δ is small compared to the irradiation spot diameter d, then after a time $t \approx \delta^2/4\nu$ the thermocapillary flow is formed:

$$V_X \approx |\partial\sigma/\partial T|\partial T/\partial r\times \delta/\eta$$

Neglecting convective heat transfer, the temperature gradient on the melt layer surface (at focusing spot radius $d/2$) can be estimated as $\partial T/\partial r \approx 2q\delta/\lambda d$, where q is the absorbed power density, λ is the thermal conductivity. For the thermocapillary flow velocity can be written

$$V_X \approx 2|\partial\sigma/\partial T|q\delta^2/(\lambda\eta d) \tag{2}$$

To estimate the velocity V_X using the expression (2), it is necessary to determine the thickness of the viscous layer δ from the flow conservation condition. The radial thermocapillary spreading of molten layer at a rate V_X under the fast laser heating leads to a lowering of liquid level at a rate V_S determined by the condition of mass conservation $V_S \approx 4V_X\,\delta/d$. The growth of velocity V_S causes in response the decreasing of molten layer thickness h and the increasing of temperature gradient $\partial T/\partial z$, which values will have quickly (during the heat propagation time $t \approx h^2/4\chi$) corrected by an increasing of melting boudary rate V_M according to the condition of heat flux conservation $q/\lambda=\partial T/\partial z$. As a result, both interfacial boundaries ("solid - liquid" and "liquid - gas") move simultaneously deep into the metal in a steady state mode, forming a deep crater. This steady-state melting process is characterized by the equality of velocities $V_M = V_S$ at a constant thickness of molten layer h and viscous layer thickness δ. Estimating the melting front velocity as $V_M \approx 2(\chi/t)^{1/2}$, and the time period needed for establishing the melt flow as $t \approx \delta^2/(4\nu)$, it can be written $V_M \approx 4(\chi\nu)^{1/2}/\delta$.

Let us write the equality of velocities $V_M = V_S$ for cavity formation during deep penetration mode:

$$(\chi\nu)^{1/2}/\delta = V_X\,\delta/d$$

Or after application (2):

$$(\chi\nu)^{1/2}/\delta = 2|\partial\sigma/\partial T|q\delta^3/(\lambda\eta d^2)$$

It turns out for an estimation of thickness of viscous thermocapillary layer during melting in deep penetration mode:

$$\delta = [(\lambda\chi^{1/2}\eta^{3/2}d^2)\ /\ (2\rho^{1/2}|\partial\sigma/\partial T|q)]^{1/4} \tag{3}$$

Having determined the viscous layer thickness δ from (3), we can then use relation (2) to determine the thermocapillary flow velocity V_X during motionless beam action depending on the irradiation conditions (beam power and the focusing spot diameter) and the cavity growth rate $V_S = 4V_X\,\delta/d$. Calculations according (3) are confirmed by direct measurements of the melt flow velocities on the front wall inside a keyhole [23] when welding of steel with a scanning beam focused into a spot with a diameter $d = 0.9$ mm. Average melt velocities $V_X = 7.5$ m/s, 10.77 m/s and 16.24 m/s were recorded At beam power of 6, 10 and 14 kW respectively. The calculated values of the thermocapillary flow velocity for these conditions with use the iron properties from Table 1 ($V_X = 10$ m/s, 13 m/s and 15.3 m/s) are very close to these empirical values. Such proximity confirm the thermocapillary mechanism of melt removal on the front wall inside a keyhole in [23].

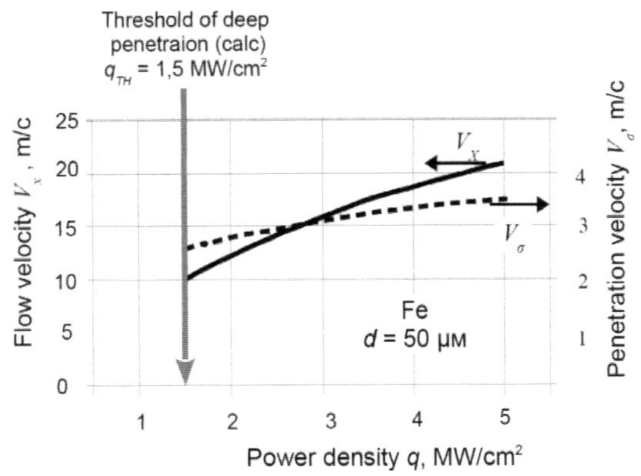

Figure 1. Change of thermocapillary melt flow velocity V_X and the rate of beam penetration (crater growth) V_S by a fixed beam ($d = 50$ μm) at SLM of iron powder layer in deep penetration mode (a little above the threshold power density $q_{TH} = 1.5$ MW/cm^2 for deep penetration).

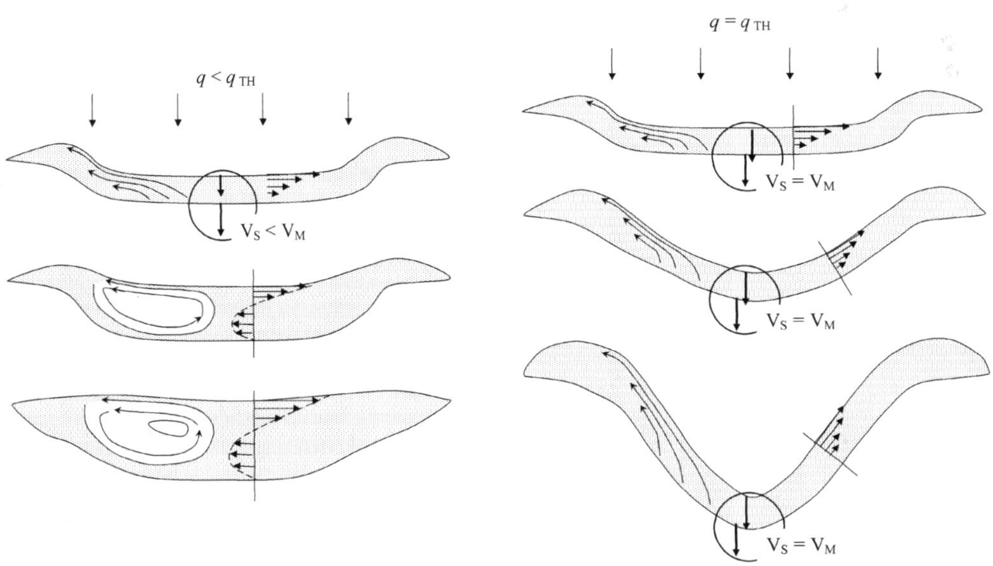

Figure 2. (a) - pre-threshold melting ($q < q_{TH}$ and $V_S < V_M$) leads to rapid flow formation with circulating structure, (b) - upon reaching the deep penetration threshold ($q > q_{TH}$ and $V_S > V_M$) the preservation of shear structure in flow leads to a keyhole formation

The obtained relations allow estimating the main hydrodynamic parameters of the thermocapillary flow in SLM technology. For example, during SLM of iron powder in deep penetration mode (d = 50 μm and beam power of 20-100 W), the temperature gradient on melt surface reaches values $dT/dr \sim 10^7$-10^8 °K/m, which gives rise to the thermocapillary melt removal from the irradiation zone with speed of $V_X \sim$ 10-20 m/s and beam penetration into powder layer with speed $V_S \sim$ 2.5 - 3.5 m/s (Fig. 1). As will be shown below (Fig. 6), the thickness of viscous layer at the same conditions is only $\delta \sim$ 1.5 - 3.5 μm. It should also be noted that the velocity values $V_X \sim$ 10-20 m/s and $V_S \sim$ 2.5 - 3.5 m/s are obtained near the threshold power density q_{TH} = 1.5 MW/cm^2 (iron, d = 50 μm), which is determined by the ratio (5), which will be discussed below.

In order to formulate the threshold condition of a deep penetration mode transition with use the thermocapillary mechanism, it is necessary to analyze the hydrodynamic processes in a thin molten layer at beginning of melting. Thermocapillary flow under the point laser impact starts with the appearance of a thin molten layer of thickness h. At the initial stage of melting, it always has the structure of shear flow with sticking of viscous layer δ to the melting boundary, as shown in the upper images Fig. 2. The further evolution of the hydrodynamic processes depends on the ratios of the velocities of the phase boundaries limiting the molten metal layer (V_M and V_S).

Fig. 2 (a) presents the evolution of the molten layer under subthreshold conditions at $q < q_{TH}$ and $V_S < V_M$. When sufficient thickness is reached in a molten layer, the viscous layer is separated from solid melting boundary ($h > \delta$), which creates favorable conditions for ring connection a current lines and the formation of circulating thermocapillary flow. It should be noted that the inequality $V_S < V_M$ become stronger at small values of q and V_X and the transition from shear to circulation structure of thermocapillary flow occurs faster than at high values of q and V_X.

The molten layer evolution at threshold power density ($q = q_{TH}$) is shown in Fig. 2 (b). A sufficiently high speed of thermocapillary flow V_X provides fast lowering of molten layer in this case. At the same time to fulfill the conditions of mass conservation and heat flux conservation the melting boundary is also going to move deeper into the metal with the same speed $V_M = V_S$, ensuring the constancy of molten layer thickness h and preserving the shear structure of flow during deep cavity formation. The longer condition $V_M = V_S$ is fulfilled and the shear structure of flow is preserved, the deeper cavity is formed.

The condition of thermocapillary "drying" $V_S > V_M$ corresponding to the exceeding of threshold power density ($q > q_{TH}$) seems to exist only as a short-term fluctuation because it must be quickly compensated by the movement of melting front to meet the heat flux conservation. We should expect an alternation of $V_S > V_M$ and $V_S < V_M$ ratios in this case with the generation of surface structures similar to those previously observed in experiments [24]. Since the existence of this fluctuation is determined by very short thermal ($h^2/4\chi$) and viscous ($\delta^2/4\nu$) response times, which are estimated as $10^{-5} \div 10^{-6}$ sec, we can assume $V_S \approx V_M$ in case of $q > q_{TH}$.

3. Conditions for sticking of viscous layer ($\delta = h$) and maintaining shear structure of thermocapillary flow during transition to deep penetration mode

The necessary condition for transition to deep penetration mode under thermocapillary mechanism of melt removal is preservation of shear structure of thermocapillary flow. Such a structure should be maintained during the time of crater formation until it reaches a maximum depth under static beam exposure. When melting by scanning beam, maintaining shear structure of thermocapillary flow is ensured by running of cold metal into beam impact zone with maintenance of thickness of molten layer within of small values close to viscous layer size ($h \sim \delta$).

The shear structure of the stream is formed simultaneously with the thermocapillary melt flow starts when the molten metal layer is still very thin. This structure is maintained as long as the thickness of the viscous layer δ remains close to the thickness of the molten layer ($\delta \sim h$) and the condition of sticking the viscous layer to the solid wall (melting boundary) is met. The deep penetration effect occurs if the shear structure of thermocapillary flux is maintained for the time needed to form a crater.

With slow heating of liquid metals, the shear structure is not maintained, as the viscous layer is separated from the solid bottom, converting the shear structure of stream into a recirculated one without forming a crater. This case (Fig. 2 (a) $q < q_{TH}$, $V_S < V_M$) is characterized by the usual Prandtl number $Pr = \nu/\chi$, which shows the ratio of temperature fields and velocity fields. If heating is slow enough the Prandtl number for melted metals is very small ($Pr = \nu/\chi << 1$) and the thickness of viscous layer is significantly smaller than the temperature propagation distance ($\delta << h$). For slow heating the condition of sticking of viscous layer to solid bottom is not fulfilled. So, the shear flow is not formed and deep penetration mode is not realized.

However, at fast heating by laser beam of high power when temperature gradient in metal reaches very high degree $\partial T/\partial z \sim 10^9$ °K/m, and the speed of a thermocapillary flow is about $V_X \sim 10$ m/s, the values of δ and h can keep commensurable magnitudes ($\delta \sim h$) during time needed for cavity formation.

This condition is disrupted upon reaching the crater to its extreme depth, when the speed of thermocapillary flow slows down and the thickness of the molten layer starts to grow rapidly, causing separation of the viscous layer separation from the solid boundary melting with subsequent closure of streamlines and transformation of shear structure of thermocapillary flow in circulation flow structure.

From this moment, the crater flows, fills with the return flow of circulation stream until it disappears completely. Such a complete and final collapse of the crater has not yet been described in studies, although it has been observed in some experiments [21, 25].

In the case of melting with a scanning beam, the crater is maintained by a stream of solid and cold metal, which rushes into irradiation zone through the front wall of keyhole and thereby provides a commensurability of δ and h.

Taking into account that effective absorption of laser radiation in keyhole close to 100% [8], we write the condition of commensurability h and δ near threshold conditions as $\delta \sim h \approx \lambda \Delta T/q$. Given that the melt temperature in keyhole is limited by the boiling temperature T_B [26] , we get:

$$\delta \sim h \approx \lambda(T_B - T_M)/q$$

Determining δ from (3), at given size of focus spot d, we obtain an expression for the threshold power density [13]:

$$q_{TH} \approx \lambda (T_B - T_M)^{4/3}(d\sigma/dT)^{1/3}(\rho/\chi)^{1/6}/(4d^{2/3}\eta^{1/2}) \quad (4)$$

or for the threshold beam power

$$P_{TH} \approx \pi \lambda[d (T_B - T_M)]^{4/3}(d\sigma/dT)^{1/3}(\rho/\chi)^{1/6}/(4\eta^{1/2}) \quad (5)$$

The resulting threshold values of power and intensity depend on the diameter of focus spot d. Let us show that the obtained dependences agree well with the experimental data.

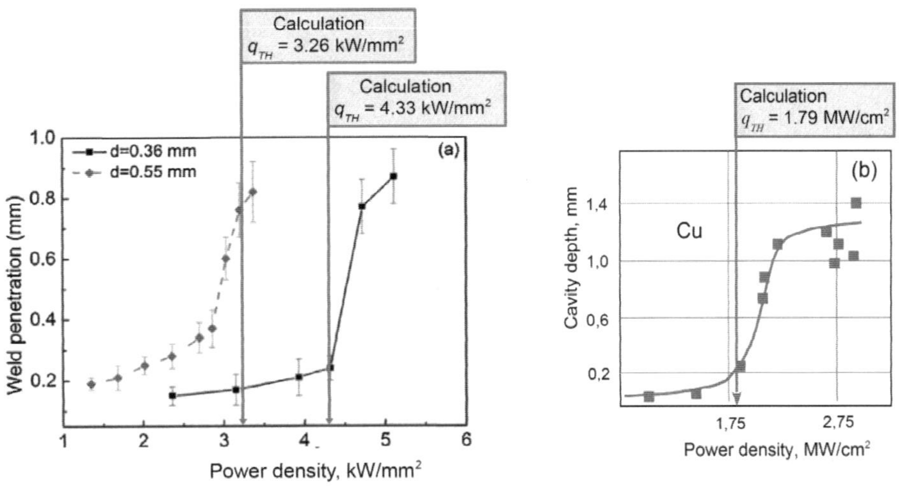

Figure 3. Transition to deep penetration (keyhole) mode in welding: (a) - steel A3 (experiment [27]), (b) - copper (experiment [28]). The arrows indicate the threshold values q_{TH} calculated with (4).

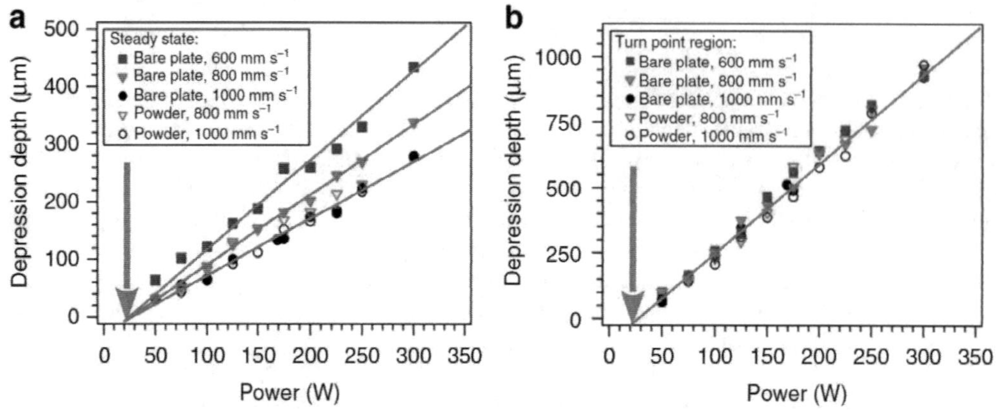

Figure 4. Surface depression depth during SLM of Ti–6Al–4V in the steady state scanning (a) and in turn point of laser beam (b) [2] in dependence of beam power and scan speed. Linear extrapolation of data gives an empirical value of threshold power for transition to deep penetration mode in experiments [2] as $P_{TH} \approx 20$ W (at the used beam with spot diameter $d = 50$ μm). The red arrows indicate the power values

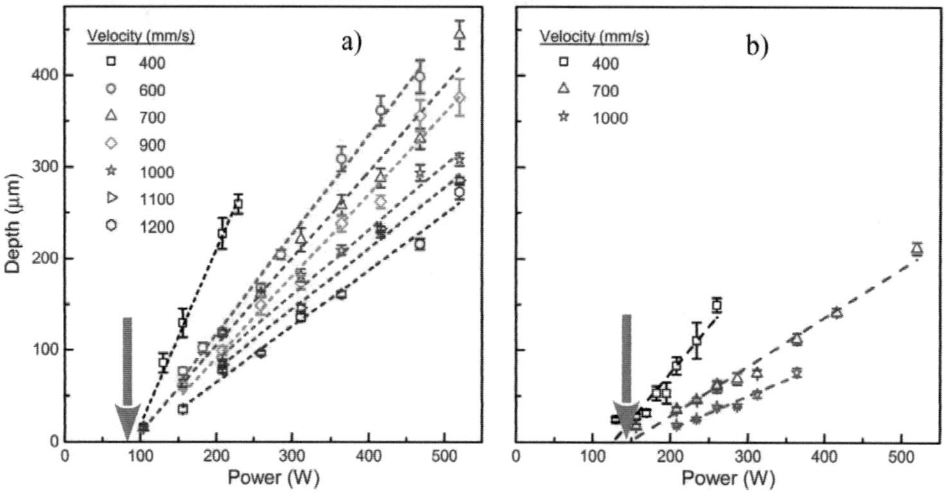

P_{TH}= 21 W calculated for $d = 50$ μm using (5).

Figure 5. Surface depression depth during SLM of Ti–6Al–4V with diameter of beam spot $d = 95$ μm (a) and $d = 140$ μm (b) [3] as function of power and scan speed. The red arrows indicate the threshold power P_{TH}=79 W calculated by (5) for $d = 95$ μm (a) and P_{TH}=143 W for $d = 140$ μm (b).

The threshold power density q_{TH} for steel A3 at two values of focal spot diameter $d = 0.36$ and $d = 0.55$ mm determined in the experiments [27], the results of which are presented dependencies of the penetration depth L from q for two values of d (Fig 3-a). The q_{TH} calculations (4) for these experiments agree well with the measurements and correctly reflect the influence of beam diameter on the threshold power density. Figure 3 (b) shows a similar comparison and good agreement of calculated value q_{TH} for copper using data from experiments [28].

Figures 4 and 5 represent an experimental data on SLM of Ti-6Al-4V at $d = 50$ μm [2], $d = 95$ μm and 140 μm [3] as a change of surface depression depth in dependence of beam power and of scanning speed. Applying linear extrapolation to the data presented in Fig. 4 (a) and (b), it is possible to determine the threshold power for the transition to deep penetration mode, which is 20 W. We will receive from (5) the calculated power value $P_{TH} = 21$ W for the conditions of experiment [2] with the properties of titanium taken from Table 1. Similarly, we obtain the corresponding calculated values for the experimental conditions [3]: P_{TH}=79 W for $d = 95$ μm (Fig. 5-a) and P_{TH}=143 W for $d = 140$ μm (Fig. 5-b) for the conditions of experiment [3].

The high correlation of the calculated and empirical parameters, which is observed in a wide range of changes in the beam parameters for different metals, convincingly confirms the thermocapillary mechanism of deep penetration effect.

Table 1. Properties of metals.

Properties	Fe	Ti	Cu
Density, кg/m^3	7800	4380	9150
Heat conductivity, W/(m×grad)	30.8	30	320
Thermal diffusivity, ×10^{-6} m^2/s	5.36	9	80
Dynamic viscosity, ×10^{-3} N×s/m^2	4.95	5.2	3.1
Temperature coefficient of surface tension, ×10^{-3} N/(m×grad)	0.49	1.075	0.21
Melting temperature, °C	1535	1680	1083
Boiling temperature, °C	3050	3300	2590

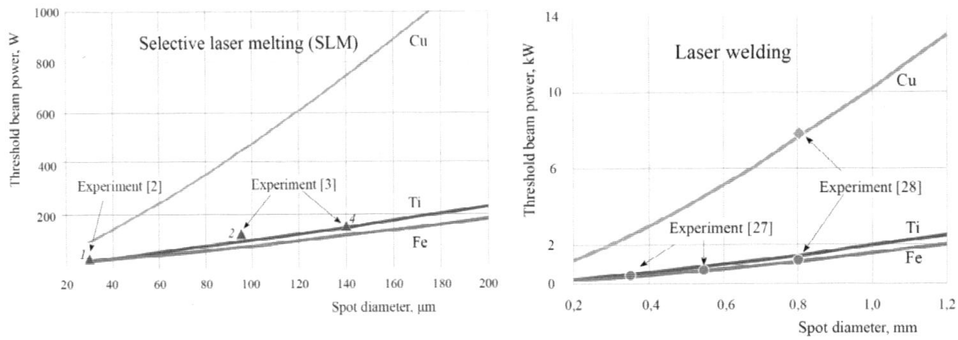

Figure 6. Conditions of threshold transition to deep penetration mode for structural metals (Cu, Fe, Ti) at change of laser power and diameter of focus spot for SLM technology (a) and for laser welding (b). Points are the experimental data for laser welding [27], [28] and SLM of Ti-6Al-4V [2], [3].

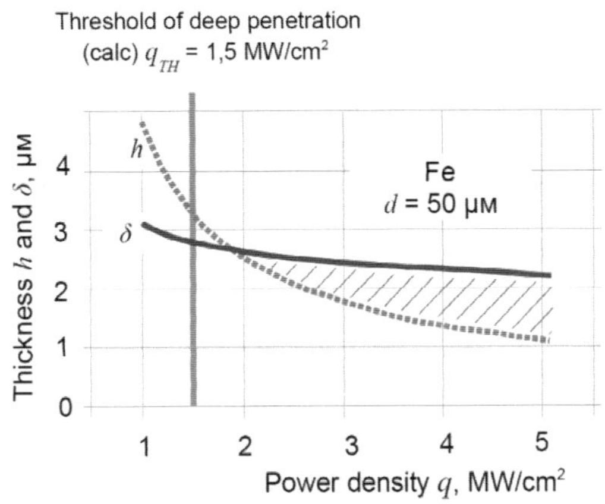

Figure 7. The fulfillment of hydrodynamic condition of thickness comparability of viscous (δ) and molten (h) layers in molten iron for SLM and beam focused in spot d = 50 μm above the calculated threshold value of beam power density is 1.5 MW/cm^2. The hatched region corresponds to the condition of sticking of viscous layer to the melting boundary ($h = \delta$).

Conditions of threshold transition to deep penetration mode for different metals (Cu, Fe, Ti) at change of laser power and diameter of focus spot are presented in Fig. 6. The lines show such a calculated (5) conditions at operating parameters of laser beam for SLM (Fig. 4, 5) and for keyhole welding (Fig. 3). Points are the experimental data obtained for different metals.

The fulfillment of hydrodynamic condition of thickness comparability of viscous (δ) and molten (h) layers in molten iron for SLM and beam focused in spot d = 50 μm is shown in Figure 7. The calculated threshold value of beam power density is 1.5 MW/cm^2. The hatched region corresponds to the condition of sticking of viscous layer to the melting boundary ($h = \delta$).

We emphasize that the threshold condition, understood on basis of the evaporative-hydrodynamic hypothesis, does not imply dependence on spot diameter d (in contrast to the dependencies (4) and (5) obtained for the thermocapillary mechanism of keyhole formation). Therefore, the noticeable influence of the beam diameter d on the threshold value of the power density q_{TH} indirectly indicates the thermocapillary mechanism of crater formation during deep penetration.

4. Conclusion

Thus, we have considered hydrodynamic aspects of deep penetration mode based on thermocapillary mechanism of melt removal from irradiation zone. Hydrodynamic characteristics of melt flow were analyzed at characteristic parameters of radiation (d = 50 μm) used for additive manufacturing by SLM method.

The rates of hydrodynamic flow of molten iron ($V_X \sim 10 \div 20$ m/s) expected in SLM of metall powder (at d = 50 μm) were estimated using the thermocapillary model of cavity formation. The validity of this forecast is confirmed by a good correlation between the calculated and empirical values of the V_X flow rate of iron during keyhole formation by a laser beam at d = 900 μm.

The analysis of the threshold transition to deep penetration mode for thermocapillary hydrodynamic mechanism of melt removal is presented. The threshold conditions for various structural metals (Cu, Fe, Ti) and in a wide range of radiation parameters, which covers the SLM of powder bed and welding is calculated. The obtained dependences are compared to experimental data, including the results taken from analysis of published experiments on SLM of a Ti-6Al-4V. It is noted that the dependence of threshold conditions on diameter of focusing spot corresponds to thermocapillary deep penetration mechanism.

The correlation of the calculated and empirical data gives the basis for conclusion about similarity of thermocapillary hydrodynamic processes in the deep penetration mode occurring during SLM of powder layer and welding.

The condition of thermocapillary threshold transition to deep penetration mode is formulated as a necessity of sticking a viscous layer to the solid melting boundary, which is characteristic of rapid heating. The fulfillment of this condition in deep penetration mode is confirmed by checking the comparability in thickness both of viscous and molten layer near the threshold conditions.

Acknowledgments

This work was supported by the Ministry of Science and Higher Education within the State assignment FSRC «Crystallography and Photonics» RAS in part of «study of the threshold conditions of thermocapillary melting in penetration (keyhole) mode for laser welding», RFBR (Project No.16-29-11743 ofi_m) in part of «investigation of the threshold conditions of deep penetration during SLM based on the thermocapillary mechanism».

References

[1] Matialainen V-P, Piili H, Salminen A, Nyrhila O. 2015 Preliminary investigation of keyhole phenomena during single layer fabrication in laser additive manufacturing of stainless steel *Physics Procedia* **78** pp 377-387

[2] Martin A A, Calta N P , Khairallah S A, et al. 2019 Dynamics of pore formation during laser powder bed fusion additive manufacturing *Nature Communications* **10** 1987

[3] Cunningham R. et al. 2019 Keyhole threshold and morphology in laser melting revealed by ultrahigh-speed x-ray imaging *Science* **363** pp 849–852

[4] Thanki A, Goossens L, Mertens R et al. 2019 Study of keyhole-porosities in selective laser melting using X-ray computed tomography. *9th Conference on Industrial Computed Tomography (iCT)* 13-15 Feb Padova Italy (iCT 2019) (https://doi.org/10.29007/3qvj)

[5] Gunenthiram V, Peyre P, Schneider M, Dal M et al. 2017 Analysis of laser–melt pool–powder bed interaction during the selective laser melting of a stainless steel *Journal of Laser Applications* **29** (2) 022303 DOI: 10.2351/1.4983259

[6] Bidare P, Bitharas I, Ward R M et al. 2018 Fluid and particle dynamics in laser powder bed fusion *Acta Materialia* **142** pp 107-120

[7] Guo Q, Zhao C, Escano L I et al. 2018 Transient dynamics of powder spattering in laser powder bed fusion additive manufacturing process revealed by in-situ high-speed high energy x-ray imaging *Acta Materialia* **151** pp 169-180

[8] Fabbro R, Dal M, Peyre P et al. 2018 Analysis and possible estimation of keyhole depths evolution, using laser operating parameters and material properties *Journal of Laser Applications* **30** No 3 032410

[9] Gladush, G. G., Smurov I. 2011 *Physics of Laser Materials Processing: Theory and Experiment* Berlin Springer-Verlag

[10] Khomenko M D, Mirzade F.Kh 2017 Parametric investigation of microstructure after laser melting of metal powder layer *Procedia Engineering* **201** pp 645–654

[11] Batanov V A, Fedorov V B 1973 Flushing Out the Liquid Phase - a New Mechanism of Producing Crater in Planar Fully Developed Evaporation of a Metallic Target by a Laser Beam. *ZhETF Pis.* **17** No 7 pp 348-351 (http://www.jetpletters.ac.ru/ps/1541/article_23562.shtml)

[12] Andrews J G, Athgey D R 1976 Hydrodynamic limit to penetration of a material by a high-power beam *J. Phys. D: Appl. Phys.* **9** No 15 pp 2181-2194

[13] Klemens P G 1976 Heat balance and flow conditions for electron beam and laser welding *J. Appl. Phys.* **47** p 2165

[14] Fabbro R, Hamadou M, Coste F 2004 Metallic vapor ejection on melt pool dynamics in deep penetration laser welding *Journal of Laser Applications* **16** No 1 pp. 16-19

[15] Courtois M, Carin M, Le Masson P, Gaied S, Balabane M 2016 Guidelines in the experimental validation of a 3D heat and fluid flow model of keyhole laser welding *J. Phys. D: Appl. Phys.* **49** 155503 (13pp)

[16] DebRoy T, David S A 1995 Physical processes in fusion welding. *Rev. Mod. Phys.* **67** No 1

[17] Seidgazov R D 2009 Verification of the melt displacement mechanism in deep penetration laser welding *Proceedings of the 4th Int. Conf. Laser Technologies in Welding and Material Processing* Ed. by B.E. Paton and V.S.Kovalenko p 62.

[18] Seidgazov R D 2019 Analysis of The Main Hydrodynamic Mechanisms in Laser Induced Keyhole Welding *The 8th Int. Conf. on Advanced Optoelectronics and Lasers (CAOL)* Bulgaria Sozopol IEEE Xplore Digital Library (in press)

[19] Seidgazov R D 2009 Thermocapillary mechanism of melt displacement during keyhole formation by the laser beam *J Physics D: Appl Phys* **42** No 17 (175501) (7 pp)

[20] Mahrle A, Beyer E 2019 Theoretical evaluation of radiation pressure magnitudes and effects in laser material processing *Physica Scripta* **94** 075004

[21] Seidgazov R D, Senatorov Yu M 1988 Thermocapillary mechanism of deep melting of materials by laser radiation *Sov. J. Quantum Electronics* **18** (3) pp 396–398

[22] Seidgazov R D 2011 Thermocapillary Mechanism of Deep Penetration in Laser Beam Welding *Mathematical Models and Computer Simulations* **3** No 2 pp 234–244

[23] Eriksson I, Powell J, Kaplan A F H 2011 Measurements of fluid flow on keyhole front during laser welding *Science and Technology of Welding and Joining* **16** (7) pp 636-641

[24] Banishev A F, Golubev V S, Khramova O D 1993 Study of the Key-Hole formation dynamics under high power laser action upon metals *Laser Phys.* **1** (6) pp 1198-1202

[25] Kayukov S V 2000 Extension of potentialities of millisecond pulsed Nd:YAG lasers in the welding technology *Quantum Electronics* **30** (11) pp 941–948

[26] Hirano K, Fabbro R, Muller M 2011 Experimental determination of temperature threshold for metal surface deformation during laser interaction on iron at atmospheric pressure *Journal of Physics D: Applied Physics* **44** (435402)

[27] Zou J L, He Y, Wu S, Xiao R S 2015 Experimental and theoretical characterization of deep penetration welding threshold induced by 1-μm laser *Applied Surface Science* 357

[28] Garaschuk V P, Velichko O A, Davydova V V 1971 Influence of the average illumination in the spot of focusing on the depth of penetration in the pulse laser welding *Avtomaticheskaya Svarka* No 5 p 31 (in Russian)

Heat treatment mode optimization of dissimilar weld joints of EP517 steel to 36NKhTYu alloy

E V Terentyev[1,2], A Yu Marchenkov[1], A L Goncharov [1], A P Sliva[1] and K T Borodavkina[1]

[1] National Research University "Moscow Power Engineering Institute", 14, Krasnokazarmennaya str., Moscow, 111250, Russia
[2] TerentyevYV@mpei.ru

Abstract. The paper presents the results of a heat treatment impact on the weld metal hardness of EP517 steel to 36NKhTYu alloy dissimilar welded joints. Aging curves for the weld metal in the temperature range from 600°C to 750°C and with an aging time of up to 64 hours are performed. Metallographic researches of weld metal and heat-affected zone (HAZ) are carried out. It was established that the greatest increase of hardness is reached after the exposure for 32 hours at 650°C and after the exposure for 16 hours at 700°C. It was found that after the aging with the specified conditions, weld metal hardness increases by 36% (60 HV) in comparison with its hardness after welding. The microstructure research showed that the decomposition fraction of supersaturated solution in HAZ of 36NKhTYu alloy significantly depends on the temperature and reaches its ultimate value after the aging at a temperature of 700°C.

1. Introduction

The industrial application of welding technologies allows combining various physical properties in welded structures made of dissimilar alloys, which are necessary to provide the defined level of structural and operational characteristics. In particular, welded structures made of dissimilar materials are used in electric machines operating at high rotation speeds under the significant mechanical loads [1, 2].

In [3, 4] the technology for the manufacture of a welded combined shroud for a high-speed rotor of an electric machine was proposed, which consists of butt-welded segments of EP517 steel and 36NKhTYu alloy (marked in accordance with Russian state standards). The chemical composition and mechanical properties of these materials are performed in tables 1 and 2.

Table 1. Chemical composition of EP517 steel and 36NKhTYu alloy (according to Russian state standards)

Material	Elements, wt. %												
	C	Si	Mn	Cr	Mo	Ni	V	P	S	W	N	Nb	Ti	Al
EP517	0.13-0.18	≤0.5	≤0.5	11.0-12.5	1.35-1.65	1.7-2.1	0.18-0.3	≤0.03	≤0.015	0.65-1.0	0.02-0.08	0.2-0.35	-	-
36NKhTYu	<0.05	0.3-0.7	0.8-1.2	11.5-13	-	35-37	-	≤0.02	≤0.02	-	-	-	2.7-3.2	0.9-1.2

EP517 ferromagnetic steel is a high-alloy martensitic steel with carbide hardening, while 36NKhTYu alloy is a paramagnetic alloy with dispersion hardening. The shroud is designed to fix permanent magnets made of rare-earth materials, and the combination of paramagnetic and ferromagnetic shroud segments is necessary to reduce non-magnetic gap on the high-speed turbo generator pole shoes and improve the operational characteristics of the electric machine [3, 4]. The rotor rotating speed reaches up to 100 000 rpm and leads to the high mechanical stresses appearance in the shroud. The problem of providing the high strength of dissimilar welded joints from these materials is complicated by the fact that after welding a supersaturated solid γ-solution is formed in the weld metal. Meanwhile, the content of elements that form the hardening phases is significantly reduced in comparison with the 36NKhTYu alloy due to its mixing with EP517 steel in the liquid state. As a result, the weld metal hardness becomes significantly lower than the hardness of the materials being welded [3-5].

Table 2. Welded materials mechanical properties

Material		R_U, MPa	$R_{0,2}$, MPa	A_5, %
EP517	Tensile test results before heat treatment [2]	987	797	7
	Requirements of standard TU 14–1–2902–80 (Normalizing 1130°C; tempering at 750°C; quenching at 1120°C (oil); tempering at 700°C)	≥1030	≥834	≥14
36NKhTYu	Tensile test results before heat treatment [2]	1237	990	10
	Requirements of standard GOST 10994–74 (Quenching at 920-950°C (water), aging at 650-670°C, aging time 2-4 h. (air cooling)	≥1030	≥635	≥14

At the same time, it is known that due to the local strengthening effect, the strength of a welded structure containing a soft layer can reach the strength of the parent metal if the width of this soft layer is less than 0.1 of the the joint thickness [6-10]. The soft interlayer in the weld joint under study includes the weld seam and HAZ. Thus, to increase the strength of the welded joint with a soft layer, it is necessary to minimize the relative width of the layer. However, the existing technologies of electron beam welding (EBW) make it possible to obtain joints with a weld width of 1 mm at least, which at a joint thickness of 4 mm in a welded shroud does not allow achieving the ultimate strength increase due to the local strengthening effect [3, 4].

Earlier, in [3, 4], in the process of EBW technology development for a combined rotor shroud made of EP517 steel and 36NKhTYu alloy, an aging with a temperature of 650°C and an aging time of 3 hours was proposed after welding for the increase of weld joint strength. The choice of the defined heat treatment mode was based on the heat treatment guidance for the 36NKhTYu alloy. The application of described aging allowed increasing the average weld hardness from 165 HV10 to 190 HV10. Since the chemical composition of the weld is significantly different from the composition of 36NKhTYu alloy, it can be assumed that there is a possibility of further increase of weld hardness and strength due to the optimization of heat treatment mode after welding. Hence, the aim of this research was to optimize the heat treatment of the obtained welded joints of EP517 steel to 36NKhTYu alloy to provide the ultimate strength properties of the weld metal.

2. Research methodology

The samples with welded joints for the researches were cut from the shroud, which was made using the EBW according to the technology described in [2, 3]. Heat treatment of welded joints samples was carried out in a Nabertherm P330 muffle furnace at heating temperatures of 600°C, 650°C, 700°C, 750°C. Aging time was varied from 7.5 minutes up to 64 hours. After each aging, the sample was cooled in the air, the surface was prepared by grinding and polishing. Then, the surface was prepared by electrochemical etching in a 10% solution of oxalic acid in water using a PoliMat2 installation for detecting the boundaries of the weld. Hardness measurements were carried out on a Wilson 432SVD hardness tester according to the Vickers method with an indentation load of 5 kg. 7-15 measurements were provided for each sample. The maximal and minimal hardness values were removed from the obtained data to exclude random factors, and the mean value and standard deviation (STD) were calculated.

Then, the sample was loaded again into a furnace heated to the appropriate temperature, and the cycle was repeated to increase the total aging time. As a result, an aging curve of the weld metal was generated for each temperature.

The preparation of microsections for metallographic researches was also performed. After a total aging time (64 hours), the cut off fragments of welded samples were pressed into a conductive compound using a SimpliMet 2000 hot-pressing machine. The samples were prepared by grinding and polishing. Electrochemical etching was also carried out in a 10% solution of oxalic acid in water. To study the microstructure of the welded joint, a Zeiss Observer Z1m optical microscope was used.

3. Research results

The results of hardness tests are shown in table 3. The aging curves are shown in Figure 1. The aging curve at a temperature of 600°C performs a gradual increase in hardness up to an aging time of 64 hours. It can be assumed that increasing the aging time will allow achieving higher values of hardness. However, such a long aging is not efficient for use in practice. At a temperature of 650°C the ultimate hardness (225 HV5) is achieved after the aging time of 32 hours. At a temperature of 700°C, the ultimate hardness (224 HV5) is achieved after the aging time of 16 hours, after which the hardness steadily decreases. At a temperature of 750°C, the ultimate hardness (195 HV5) is achieved after the aging time of 8 hours. Moreover, its hardness decreases up to 190 HV5 with the further aging time increase from 8 hours till 64 hours.

Table 3. Hardness test results of weld metal after aging

Aging time	Aging temperature							
	600°C		650°C		700°C		750°C	
	HV5	STD	HV5	STD	HV5	STD	HV5	STD
0	**165**	5.6	**168**	3.7	**175**	4.0	**167**	7.2
7.5 min	**171**	6.7	**175**	6.5	**184**	13.0	**178**	9.3
22.5 min	**169**	4.5	**175**	5.7	**187**	12.6	**181**	11.9
52.5 min	**176**	15.4	**181**	13.4	**196**	18.0	**178**	4.0
1 h 52.5 min	**179**	16.5	**192**	12.3	**206**	26.9	**184**	10.4
4 h	**187**	21.1	**197**	14.6	**207**	33.1	**188**	14.8
8 h	**196**	29.0	**211**	16.2	**215**	28.8	**195**	19.8
16 h	**200**	28.9	**216**	13.0	**224**	44.7	**194**	19.6
32 h	**199**	30.7	**225**	13.5	**221**	40.4	**192**	20.8
64 h	**206**	33.2	**223**	17.4	**218**	26.1	**188**	13.9

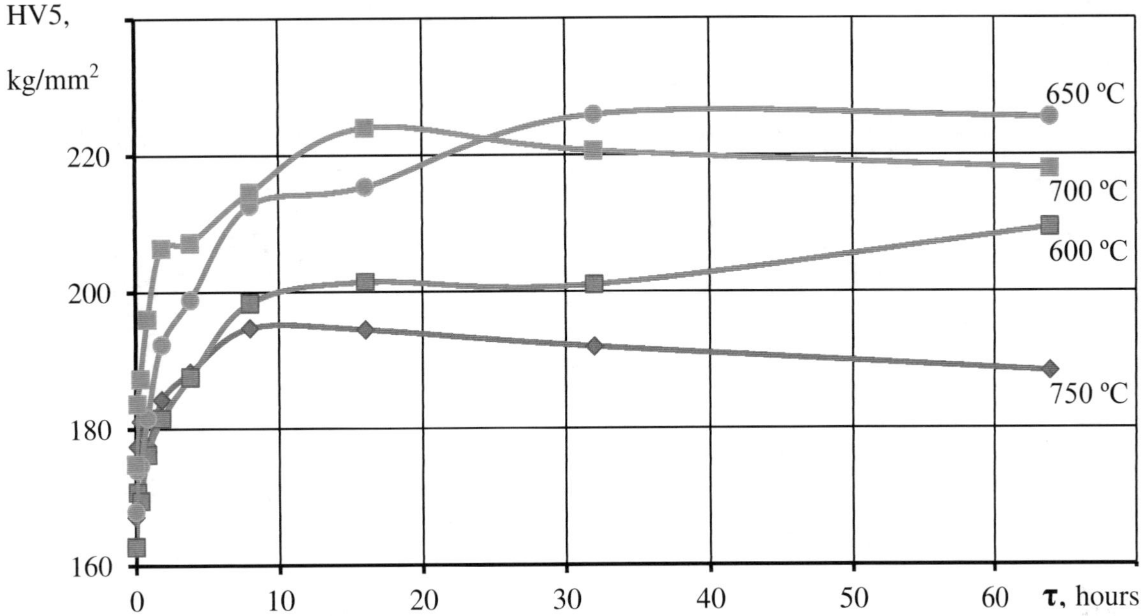

Figure 1. Aging curves of dissimilar welded joints of EP517 steel with 36NKhTYu alloy.

The study of the welded joint microstructure after aging time of 64 hours showed that in the HAZ of EP517 steel, the quenching structures decomposed with the formation of a dispersed mechanical mixture. Herewith, no any significant differences from the previously obtained results [5] were occurred.

Figure 2 shows the microstructure of weld metal and 36NKhTYu alloy HAZ after aging at various temperatures. After etching the intermetallic phases in the form of dark regions are observed in the HAZ, which were formed according by an intermittent mechanism. It can be seen that the decomposition fraction of the supersaturated solution by the intermittent mechanism in the HAZ significantly depends on temperature and reaches its maximum value after aging at a temperature of 700°C (Figure 2c). The applied research technique does not allow identifying the hardening phases that were dropped by the continuous mechanism, and these areas remain bright (Figure 2). Table 4 shows the results of hardness measuring in the areas of intermittent decomposition and release of hardening phases in the 36NKhTYu alloy HAZ. It is seen that the hardness in researched areas differs significantly, especially at an aging temperature of 600°C.

The microstructure of the weld metal after aging at different temperatures has no significant differences, which is explained by the continuous mechanism of supersaturated solid solution decomposition (Figure 2). As well as in [3, 4], insignificant regions of hardening phases intermittent drop were revealed in the weld metal, which may occur due to local chemical inhomogeneity.

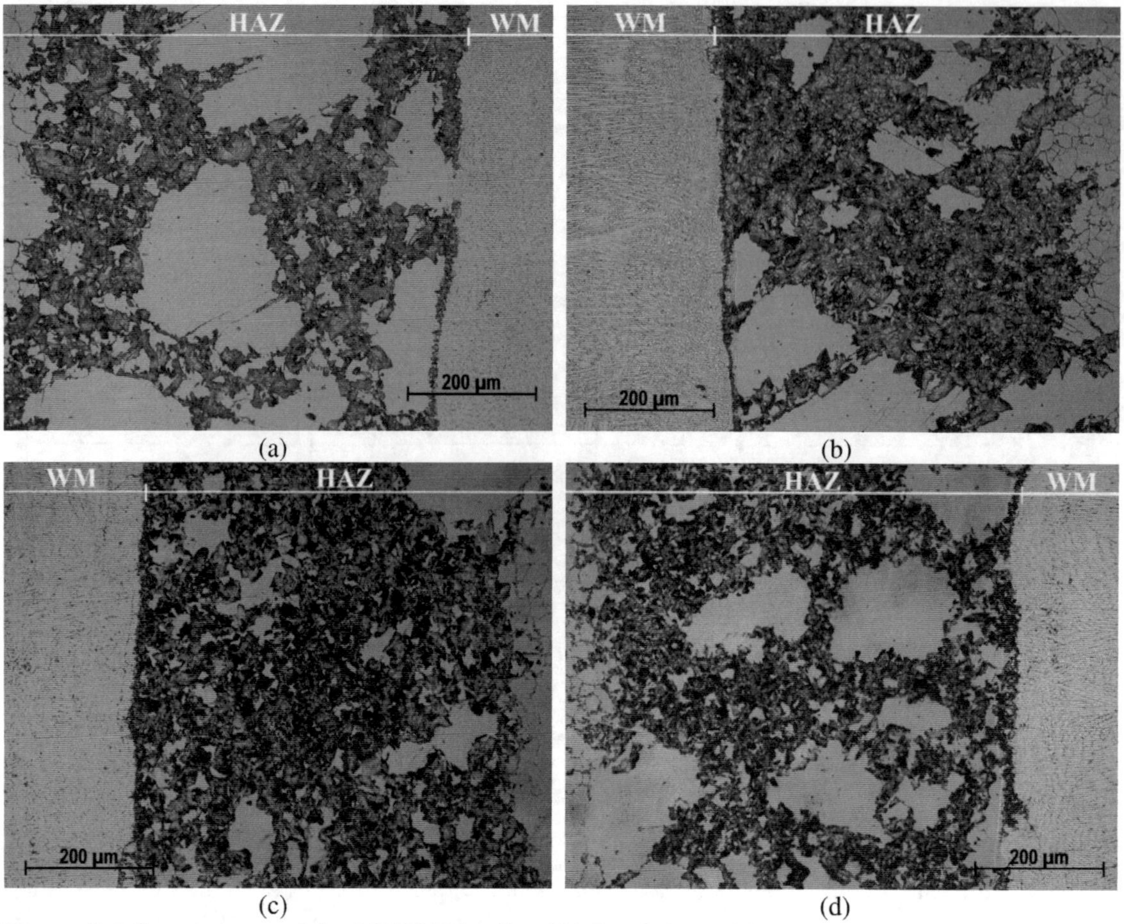

Figure 2. Microstructure of the 36NKhTYu alloy HAZ and the weld metal (WM) after aging time of 64 hours at temperatures: (a) - 600ºC; (b) - 650ºC; (c) - 700ºC; (d) - 750ºC.

Table 4. Results of the hardness measuring in the area of intermittent decomposition and in the area of continuous drop of hardening phases in the 36NKhTYu alloy HAZ

Area	Aging temperature							
	600ºC		650ºC		700ºC		750ºC	
	HV1	STD	HV1	STD	HV1	STD	HV1	STD
Intermittent decomposition area	**390**	13.7	**404**	8.5	**368**	6.4	**366**	18.1
Area of continuous drop of hardening phases	**299**	9.6	**385**	10.1	**427**	9.7	**385**	7.5

4. Discussion

Aging curves analysis shows that the maximum hardness of the weld metal can be achieved with the following aging parameters: 16 hours of aging at a temperature of 700°C or 32 hours at 650°C. Obviously, an increase in the aging temperature to 750ºC leads to over aging, i.e. decrease in hardness due to coagulation of the hardening phase. Thus, by optimizing the aging conditions of dissimilar welded joints of EP517 steel with 36NKhTYu alloy, it is possible to increase weld metal hardness by 36% (60 HV) in comparison with its hardness after welding and by 18% (35 HV) in comparison with its hardness after the heat treatment (3 hours at 650 ºC) proposed in [3-5].

It should be noted that an increase in aging time with all temperatures causes an increase in standard deviations (STD), which, presumably, can be related to heterogeneity of the weld metal. The concentration of components which form hardening phases during aging varies significantly in weld metal volume that causes the irregularity in the increase of its hardness during aging.

It is known [11] that heat treatment can lead to the hardening intermetallic phases drop by two competing mechanisms: intermittent and continuous. The occurrence of intermittent decomposition is detected structurally (Fig. 2), while continuous decomposition is indirectly confirmed by an increase in hardness of grains which are free from intermittent decomposition. The increase in the hardness of the weld metal and a change in the microstructure of 36NKhTYu alloy HAZ indicate that the initial solution was supersaturated immediately after the EBW. As it was already shown in [3, 4], intermittent decomposition is also observed in the weld metal, however, the proportion of the intermetallic phase in the weld metal formed by this mechanism is significantly less than in the HAZ. Obviously, it is caused by significant differences in the chemical composition and, in particular, by different contents of the elements that form the hardening phases during aging. In order to identify the mechanisms of the weld metal hardening, further investigation of the microstructure changes features during aging using the electron microscopy with high resolution is required. It is worth noting that in order to assess the structural strength of the welded structure after aging, it is necessary to study the strength characteristics of all weld joint zones, for example the heat-affected zones of EP517 steel and 36NKhTYu alloy and the parent metal, where structural changes can be also occurred.

5. Conclusion

- The maximum increase of the weld metal hardness of EP517 steel to 36NKhTYu alloy weld joint can be achieved with the following aging parameters: 16 hours of aging at a temperature of 700°C or 32 hours of aging at a temperature of 650°C. The hardness increases by 36% (60 HV) in comparison with the weld metal hardness after welding.
- The increase of aging time leads to an increase in the hardness values range because of the structural and chemical heterogeneity of the weld metal.

Acknowledgements
The study was performed at the Moscow Power Engineering Institute and was supported by the Russian Science Foundation (project No. 18-79-00335).

References

[1] Bespalov V Ya 2006 *Electric machines* (M.: “Academy”)

[2] Dragunov V K, Goncharov A L, Terentyev E V, Marchenkov A Yu 2015 *Creation of welded combined structures in power engineering. Part 1: Physical processes in the welding of dissimilar metals* (M.: “Veche)

[3] Terentyev E V, Marchenkov A Yu, Goncharov A L, Sliva A P 2018 Improving the structural strength of dissimilar welded joints of 36NKhTYu alloy and EP517 steel by optimizing the geometric parameters of the weld *Vestnik Permskogo natsional'nogo issledovatel'skogo politekhnicheskogo universiteta. Mashinostroyeniye, materialovedeniye* **20(3)** 63-72

[4] Dragunov V K, Sliva A P, Terentyev E V, Goncharov A L , Marchenkov A Yu, Portnov M A 2018 EBW technology of combined bandage of high-speed electric machine rotor *13th international conference on electron beam welding “E+E”* (vol 53 no 5-6) pp 112-118

[5] Terentyev E V, Marchenkov A Yu, Goncharov A L, Sliva A P 2019 Application of Local Strengthening for Increasing the Strength of the Welded Joint of EP517 Steel and 36NKhTYu Alloy *Russian Metallurgy (Metally)* **2019(10)** 1061–1066

[6] Erofeev V V, Ignatiev A G 2015 Justification of the electron beam welding modes of membrane-type products from AMg6N and AMg6NPP alloys *Vestnik YUUrGU Seriya «Mashinostroyeniye»* **15(4)** 53-61

[7] Shakhmatov M V, Erofeev V V, Kovalenko V V 1999 *Production technology and calculation of welded membranes* (Ufa: poligrafkombinat)

[8] Bakshi O A, Erofeev V V, Anisimov Yu I, Shakhmatov M V, Yaroslavtsev S I 1983 The effect of the mechanical heterogeneity degree on the static strength of welded joints *Svarochnoye proizvodstvo* **4** 1
[9] Dilman V L, Ostsemin A A, Eroshkina T V 2008 Strength of mechanically dissimilar welded joints of reinforcing bars *Vestnik mashinostroyeniya* **9** 13-16
[10] Dilman V L, Ostsemin A A 2005 The stress state and static strength of the plastic layer during plane deformation *Problemy mashinostroyeniya i nadezhnosti mashin* **4** 38-48
[11] Novikov I I 1986 *Theory of metals heat treatment. Textbook. 4th ed., Rev. and add* (M: Metallurgy)

AUTHOR INDEX

Abusaif, N ... 106
Agafonov, R Yu ... 66, 86
Antonovich, D A ... 1
Balanovsky, A E ... 8, 60, 72
Belenkiy, V Ya ... 99
Beniyash, A ... 13
Bernard, Alain ... 77
Biryukov, V P ... 23
Blokhina, K A ... 29
Borodavkina, K T ... 160
Chulkov, I S ... 37
Chulkova, A V ... 37, 53
Demidov, A N ... 106
Devyataykina, T A ... 29
Dmitrenko, D V ... 126
Drimal, D ... 43
Fedoseeva, E M ... 132
Gaponova, D A ... 144
Geraskin, E I ... 106
Goncharov, A L ... 37, 53, 160
Gozbenko, V E ... 8, 60, 72
Grigoriev, G V ... 66, 86
Gruzdev, V A ... 1
Gudenko, A V ... 66
Hassel, T ... 13
Kapil, Sajan ... 77
Kargapol'Tsev, S K ... 8, 60, 72
Karimbekov, M A ... 106
Karlina, A I ... 8, 60, 72
Karlina, Yu I ... 8, 60, 72
Karunakaran, K P ... 77
Kasitsyn, A N ... 66, 86
Kharitonov, I A ... 53
Kisarev, A V ... 92
Klimov, G ... 13
Kobernik, N V ... 92
Kolenic, F ... 43
Koleva, E G ... 139
Leonov, A V ... 86
Letyagin, I Yu ... 99
Marchenkov, A Yu ... 106, 160
Matyunin, V M ... 106
Medvedev, A M ... 29
Milonov, A S ... 112
Mindibaev, M K ... 139
Mirzade, F Kh ... 150
Mishigdorzhiyn, U L ... 112
Moskvitin, G V ... 119
Nambolan, Arun ... 77
Nazarko, A S ... 126
Negi, Seema ... 77
Olshanskaya, T V ... 132
Pandya, Milan ... 77
Pang, S ... 99
Panov, D V ... 23
Patel, Ashik ... 77
Patil, Yogesh ... 77
Permyakov, G L ... 139
Petrovsky, V N ... 23
Plomoyalo, R L ... 126
Pugachev, M S ... 119
Rodyakina, R V ... 144
Sealy, Michael ... 77
Seidgazov, R D ... 150
Semenov, A P ... 112
Semenov, Yu I ... 29
Shcherbakov, A V ... 144
Simek, M ... 43
Sizov, M M ... 29
Sliva, A P ... 53, 66, 160
Starostenko, A A ... 29
Terentyev, E V ... 53, 160
Trushnikov, D N ... 99, 139
Trushnikov, Dmitriy ... 77
Tsyganov, A S ... 29
Ulakhanov, N S ... 112
Ushakov, D V ... 23
Volkov, P V ... 106
Zhgut, D A ... 106